ICE AGES

ICE AGES
SOLVING THE MYSTERY

John Imbrie and
Katherine Palmer Imbrie

Enslow Publishers
Short Hills, New Jersey 07078

QE
697
.I 45

Library of Congress Cataloging in Publication Data

Imbrie, John.
  Ice ages.

  Bibliography: p.
  Includes index.
  1. Glacial epoch. 2. Glaciology. I. Imbrie,
Katherine Palmer, joint author. II. Title.

QE697.I45   551.7'92   78-13246

ISBN 0-89490-015-3
ISBN 0-89490-020-X pbk.

Printed in the United States of America

10   9   8   7   6   5   4   3   2   1

CONTENTS

PREFACE   7
LIST OF ILLUSTRATIONS   9
PROLOGUE: The Forgotten Ice Age   11

PART I: ICE AGE DISCOVERED

1: Louis Agassiz and the Glacial Theory   19
2: The Triumph of the Glacial Theory   33
3: Exploring the Ice-Age World   47

PART II: EXPLAINING THE ICE AGES

4: The Ice-Age Problem   61
5: Birth of the Astronomical Theory   69
6: The Astronomical Theory of James Croll   77
7: Debate Over Croll's Theory   89
8: Through Distant Worlds and Times   97
9: The Milankovitch Controversy   113
10: The Deep and the Past   123
11: Pleistocene Temperatures   135
12: Milankovitch Revival   141
13: Signal from the Earth   147
14: Pulsebeat of Climate   153
15: Pacemaker of the Ice Ages   161

PART III: ICE AGES OF THE FUTURE

16: The Coming Ice Age   177
EPILOGUE: The Last Billion Years of Climate   189
APPENDIX: Chronology of Discovery   195
SUGGESTED READING   203
BIBLIOGRAPHY   205
INDEX   215

PREFACE

Knowledge that the world once endured an ice age has been widespread for more than a century. Indeed, the concept is now so familiar that nearly every winter storm prompts dramatic headlines: Is a new ice age upon us?

This book tells the story of the ice ages—what they were like, why they occurred, and when the next one is due. It is a tale of scientific discovery, and therefore a book about people—about the astronomers, geochemists, geologists, paleontologists, and geophysicists from a dozen countries who have been engaged for nearly a century and a half in the search for a solution to the ice-age mystery.

We are indebted to many people for their advice and assistance in the preparation of this book. In particular we wish to thank Vasko Milankovitch for sharing with us his memories of his father, Milutin Milankovitch. Tatomir P. Angelitch of the Serbian Academy of Sciences and Arts was kind enough to provide us with a complete list of Milankovitch's publications. Barbara Grönquist translated many passages from the German. Gordon Craig of the University of Edinburgh helped locate information about James Croll. The studies by Albert V. Carozzi of the University of Illinois were an invaluable guide to the early development of the glacial theory. Catherine Krause located obscure sources that contributed greatly to this book.

Many people put their personal recollections at our disposal. Among them were William A. Berggren, Wallace S. Broecker, Rose Marie Cline, Cesare Emiliani, Samuel Epstein, David B. Ericson, Rhodes W. Fairbridge, James D. Hays, George J. Kukla, Robley K. Matthews, Neil D. Opdyke, Nicholas J. Shackleton, and Manik Talwani.

Rosalind M. Mellor prepared the typescript and kept an eagle eye out for inconsistencies. Terry A. Peters also helped in this preparation. Barbara Z. Imbrie gave the manuscript its first critical reading and made many helpful suggestions.

The Museum of Science in Boston was kind enough to grant one of us (K.P.I.) a leave of absence so that this book might be written.

Finally, we wish to express our gratitude to our editor, Ridley Enslow, for his encouragement and advice. In large part, his enthusiasm and interest have made this book a reality.

Seekonk, Massachusetts   J.I.
June, 1978   K.P.I.

LIST OF ILLUSTRATIONS

1. Earth today and during the last ice age.   12, 13
2. Glacial deposit on Cape Ann, Massachusetts.   15
3. Scratched stone from a glacial deposit in Europe.   20
4. Zermatt glacier in the Swiss Alps.   23
5. Portrait of Louis Agassiz at the Unteraar Glacier.   27
6. Polished bedrock near Neuchâtel, Switzerland.   29
7. Erratic boulder in Scotland.   38
8. The Reverend Professor Buckland, equipped as a "glacialist."   42
9. Antarctic Ice Sheet.   44
10. Chamberlin's map of North America during the ice age.   50
11. Shorelines of ancient Lake Bonneville, Utah.   56
12. Multiple tills in Scotland.   57
13. March of the seasons.   69
14. Dates of equinox and solstice.   70
15. Precession of the earth.   73
16. Precession of the equinoxes.   74
17. Ellipses with different eccentricities.   82
18. Orbital eccentricities calculated by James Croll.   82
19. Croll's theory of the ice ages.   84
20. Photograph of James Croll.   87
21. Succession of fossiliferous strata according to Charles Lyell.   90
22. Lyell's classification of earth history.   91
23. Modern classification of the Cenozoic Era.   92
24. Milankovitch radiation curve for latitude 65° North.   105
25. Effect of axial tilt on the distribution of sunlight.   107
26. Milankovitch radiation curves for different latitudes.   108
27. Milutin Milankovitch.   110
28. Theoretical succession of North American ice ages.   115

29. Theoretical succession of European ice ages.   116
30. Eberl's test of the Milankovitch theory.   118
31. Fluctuations of the ice-sheet margin
     between Indiana and Quebec.   122
32. Fossil from the deep-sea floor.   130
33. Succession of Caribbean ice ages
     according to Ericson and Emiliani.   132
34. Reef terraces on New Guinea.   145
35. Astronomical theory of Barbados sea levels.   146
36. Magnetic history of the earth.   150
37. Climatic history recorded in a
     Czechoslovakian brickyard.   155
38. The 100,000-year pulse of climate.   157
39. The "Rosetta Stone" of late Pleistocene climate.   165
40. Climate of the past half-million years.   169
41. Changes in eccentricity, tilt, and precession.   170
42. Spectrum of climatic variation over the past
     half-million years.   171
43. Climate of the pastyears.   179
44. Climate of the past 100 years.   180
45. Climate of the past 1,000 years.   181
46. The Argentière glacier today and in 1850.   182
47. Climatic forecast to the year A.D. 2100.   185
48. Climatic forecast of the next 25,000 years.   186
49. The last billion years of climate.   190

The Forgotten
Ice Age

Twenty-thousand years ago, the earth was held in thrall by re-
lentlessly probing fingers of ice—ice that drew its power from
frigid strongholds in the north, and flowed southward to bury
forests, fields, and mountains. Landscapes that were violated by
the slowly moving glaciers would carry the scars of this advance
far into the future. Temperatures plummeted, and land surfaces
in many parts of the world were depressed by the unrelenting
weight of the thrusting ice. At the same time, so much water was
drawn from the ocean to form these gargantuan glaciers that sea
levels around the world fell by 350 feet, and large areas of the
continental shelf became dry land.

This period in the earth's history has come to be called the ice
age. In North America, glacial ice spread out from centers near
Hudson Bay to bury all of eastern Canada, New England, and
much of the Midwest under a sheet of ice that averaged more than
a mile in thickness. A second ice sheet spread out from centers in
the Canadian Rockies and other highlands in western North
America to engulf parts of Alaska, all of western Canada, and
portions of Washington, Idaho, and Montana. In Europe, the ice
reached outward from Scandinavia and Scotland to cover most of
Great Britain, Denmark, and large parts of northern Germany,
Poland, and the Soviet Union. A smaller ice cap, centered on the
Alps, buried all of Switzerland and nearby portions of Austria,
Italy, France, and Germany. In the southern hemisphere, small
ice sheets developed over parts of Australia, New Zealand, and
Argentina. In all, the ice covered about 11 million square miles of
land that is today free of ice.

Immediately south of these great northern-hemisphere ice
sheets, the landscape was treeless tundra. Here, during the short,

Figure 1.   Earth today (left) and during the last ice age (right). Twenty-
thousand years ago, great ice sheets covered parts of North America,
Europe, and Asia; surface waters of the Arctic and parts of the North
Atlantic Oceans were frozen; and sea level was 350 feet lower than it is

today. Many parts of the continental shelf, including a corridor between Asia and North America, became dry land. (Drawing by Anastasia Sotiropoulos, based on information compiled by George Denton and other members of the CLIMAP project.)

cool summers, heather and other hardy, low-growing plants grew in the boggy soil. Migrating herds of reindeer and mammoths grazed upon this lush plant cover during the summer months, and in winter moved southward seeking more favorable pastures. In North America, the tundra was only a narrow belt of land that served to separate the ice sheets to the north from forested areas to the south. In the eastern part of the continent, spruce trees grew in a continuous forest; in the more arid Midwest, the stands of spruce followed the rivers, while in between there were dusty grasslands.

In Europe and Asia, the tundra belt was wider than it was in North America, giving way only gradually to a vast expanse of semiarid grassland that stretched from horizon to horizon— across two continents from the Atlantic coast of France, through central Europe, to eastern Siberia.

Stone Age hunters, following herds of mammoth and reindeer across the tundra, could glimpse the southern edge of the ice sheet. As the cold penetrated their deerhide clothing, and the wind from the north whipped their faces, it would have been difficult for these people to realize that their descendants would inhabit a very different world from their own.

Yet the ice age did come to an end. About 14,000 years ago, the ice sheets began to retreat. Within 7,000 years they had withdrawn to their present limits. Today, all that remains of the ice sheets in the northern hemisphere are the Greenland Ice Sheet and a few small ice caps in the Canadian Arctic. Where modern farmers reap Iowa corn and Dakota wheat, mile-high glaciers once ground their way over the land. And, where European forests stand today, treeless plains once stretched to the horizon.

As the glaciers melted back, the landscape they left behind was greatly altered—a landscape strewn with traces of its glacial origin. In northern regions, the ice sheets had ground away at the underlying surface, scratching deep grooves in the bedrock, and swallowing bits and pieces of eroded material. This material had been transported outward to the margins of the ice sheets where it was deposited in a chaotic jumble known as a moraine (Figure 2).

As the ice sheets withdrew, human recollections of them began to fade. Racial memory—if it exists at all—must be imperfect, for the world of the Stone Age hunters was soon forgotten. Even the

Figure 2. Glacial deposit on Cape Ann, Massachusetts: the landscape is typical of areas once covered by ice sheets. (From J.D. Dana, 1894.)

clues left behind by the great ice sheets were misinterpreted. By the eighteenth century, geologists surmised that the blanket of glacial sediments had been transported and deposited by the great flood described in the Bible. It was only in the early years of the nineteenth century that some scientists began to question this explanation. Were floodwaters—even divinely inspired ones—actually capable of transporting gigantic boulders hundreds of miles, or was some other agent responsible?

PART I

ICE AGE
DISCOVERED

1

Louis Agassiz
and the Glacial Theory

Few residents of the town of Neuchâtel in Switzerland were stirring at 4:15 A.M. on the morning of July 26, 1837. Had they been, they would have observed a long line of well-made carriages creaking slowly through the cobbled streets of their sleepy town. In fact, very few people were aware that three of the most respected scientists of the day shared the first and grandest carriage, which was drawn by four white horses.

Leopold von Buch—whose gray locks and bent frame belied his boundless energy—stared morosely at the floor of the swaying carriage. Jean Baptiste Elie de Beaumont—erect and outfitted with consummate good taste despite the ungodly hour at which his valet had awakened him—glared coldly at the ice-capped peaks of the Alps, 50 miles distant across the Swiss plain, and at the surrounding, and less daunting, Jura. The third passenger in the carriage—a dark-haired, broad-shouldered young man with a bright, curious gaze—looked out of the window and reflected grimly that only Elie de Beaumont's manner could rival the chilly remoteness of the Alpine peaks that seemed to hold themselves proudly aloof from the caravan of rattling carriages.

Elie de Beaumont's frigid demeanor perturbed the young man. For Louis Agassiz, with his quick, inquisitive mind, found it incomprehensible that a scientist of Elie de Beaumont's mark could fail to see the import of this particular journey through the Jura mountains.

Two days earlier, the Swiss Society of Natural Sciences had held its annual meeting in Neuchâtel, and the young president of the society, Louis Agassiz, had startled his learned associates by presenting a paper dealing not, as the distinguished members of the society had expected, with the fossil fishes lately found in far-off

Brazil, but with the scratched and faceted boulders that dotted the Jura mountains around Neuchâtel itself (Figure 3). Agassiz argued that these erratic boulders (chunks of rock appearing helter-skelter in locations far removed from their areas of origin) could only be interpreted as evidences of past glaciation—and an ancient age of ice.

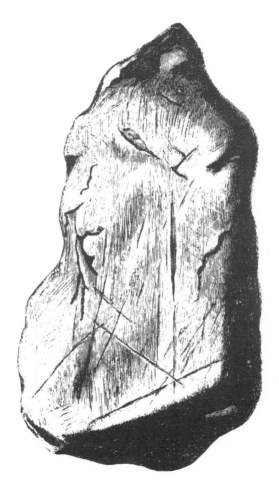

Figure 3.   Scratched stone from a glacial deposit in Europe. Boulders of this kind are common features of glaciated landscapes. (From J. Geikie, 1877.)

Thus began a dispute—one of the most violent in the history of geology—that was to rage on for more than a quarter of a century and end with universal acceptance of the ice-age theory. Although the concept of an ice age did not begin with Agassiz, his controversial paper (later known as "the Discourse of Neuchâtel") served to bring the glacial theory out of scientific obscurity and into the public eye.

As president of a Swiss scientific society, Agassiz found himself in an ideal position to present his theory to the elite of the nineteenth-century scientific world. He was, however, only one link in a chain that eventually would lead to general acceptance of the startling theory that a moving sea of ice had at one time covered large areas of the globe.

This theory, initially rejected by the most renowned scientists of the age, had long been accepted as fact by many Swiss who lived and worked in the mountains and so came into daily contact with evidence of an extensive past glaciation. A few noted scientists and naturalists became early converts to the theory, but these few lacked opportunities to promote their ideas successfully.

As early as 1787, Bernard Friederich Kuhn, a Swiss minister, interpreted local erratic boulders as evidence of ancient glaciation. Seven years later, James Hutton—the Scottish geologist who is now regarded by many to be the father of the science of geology—visited the Jura and reached the same conclusions as Kuhn. In 1824, Jens Esmark saw evidence of former extensive glaciers in Norway. Esmark's views were known to Reinhard Bernhardi, a German professor of natural science who later made observations of his own. In 1832, Bernhardi published an article arguing that a polar ice cap had once spread across Europe to reach as far south as central Germany.

Many of these early pioneers developed their ideas completely independently, through personal observation and deduction. But so deeply entrenched was the accepted explanation of erratics as the deposit of a great flood, that none of these men was able to make their revolutionary ideas widely known. It would demand the combined efforts of some of the greatest scientific minds of the age, over a period of 25 years, to overthrow the established theory.

It is not surprising that in such a religious age scientists and laymen alike believed that these boulders had been transported by unimaginably huge currents of water and mud deriving from the biblical deluge of Noah's time. This theory did undergo some revision, however, and when Agassiz made his presentation to the Swiss Society of Natural Sciences in 1837, the accepted explanation for erratics was one that had been developed in 1833 by the great English geologist, Charles Lyell. Lyell argued that the agents responsible were boulder-laden icebergs and ice rafts that had drifted about in the great flood.

The chain of creative thinkers and fortunate circumstances which led up to Agassiz' presentation at Neuchâtel began with Jean-Pierre Perraudin, a mountaineer from the southern Swiss Alps. Perraudin made his living hunting chamois near Lourtier in the Val de Bagnes. As a result of his own observations, he came to the conclusion as early as 1815 that the glaciers, which then occupied only the higher, southern portion of the Val de Bagnes, had once filled the entire valley. He wrote:

> Having long ago observed marks or scars occurring on hard rocks which do not weather, I finally decided, after going near the glaciers, that they had been made by the pressure or weight of these masses, of which I find traces at least as far as Champsec. This makes me think that glaciers filled in the past the entire Val de Bagnes, and I am ready to demonstrate this fact to incredulous people by the obvious proof of comparing these marks with those uncovered by glaciers at present.

In 1815, Perraudin communicated his ideas to Jean de Charpentier, a naturalist who would later become an important advocate of the glacial theory. Impressed by the mountaineer's observations—but as yet unconvinced—de Charpentier wrote:

> Although Perraudin extended his glacier only [24 miles beyond its present limit to Martigny], because he himself probably had never been beyond that town, and although I agreed with him on the impossibility of transporting erratic boulders by water, I nevertheless found his hypothesis so extraordinary and even so extravagant that I considered it as not worth examining or even considering.

Sometime during the next three years, however, Perraudin was to find a sympathetic ear at last in the person of Ignace Venetz, who was by profession a highway and bridge engineer. From

Figure 4. Zermatt glacier in the Swiss Alps as depicted in an illustration published by Louis Agassiz in 1840. (From A.V. Carozzi, 1967, with permission of A.V. Carozzi and the University of Neuchâtel.)

1815 to 1818, Venetz spent a great deal of time in the Val de Bagnes area in connection with his work. During this period he had many discussions with Jean-Pierre Perraudin on the subject of glaciers. It was an historic and fortunate association.

Venetz was somewhat slow to adopt Perraudin's theories. In 1816, at the annual meeting of the Swiss Society of Natural Sciences at Bern, he brought up the subject of glaciers, but only to present some ideas on glacial movement and to describe how elongate accumulations of rock debris (moraines) form along the surface of glaciers (Figure 4). Five years later, Venetz was still reluctant to commit himself fully to the glacial theory. In a memoir dated 1821 (unpublished until 1833), he noted that he had identified several ridges of rock debris located three miles beyond the terminus of the Flesch glacier and suggested that these deposits were moraines left by that glacier in earlier times.

Venetz did not fully develop the ideas he had gleaned from Perraudin until 1829 when at the annual meeting of the Society at the Hospice of the Great St. Bernard, he stated his conclusion that immense glaciers had once spread out from the Alps to cover not only the Swiss plain and the Jura, but other parts of Europe as well. In support of this theory he described the distribution of erratic boulders and moraines, and compared these deposits to those formed by contemporary Alpine glaciers.

Despite Venetz's forthright but long-delayed presentation, the theory was generally ignored or rejected outright by the scientists who were present at the meeting. One man, however, saw the truth of the theory, and subsequently came to Venetz's aid. That man was Jean de Charpentier, who had long been acquainted with Ignace Venetz. As director of the salt mines at Bex in Switzerland, de Charpentier took a lively interest in science and the natural world. Now—like the prodigal son—he was ready to back Venetz and to stand squarely behind the theory he had rejected almost 15 years earlier.

Over the course of the next five years (1829–1833), de Charpentier brought his remarkable powers of reasoning to bear on the problem of ancient glaciation. Although Venetz had been the first to accept the radically new ideas of Perraudin, it was de Charpentier who gave the glacial theory an unshakeable foothold in scientific fact by organizing and classifying the evidence that supported it. But de Charpentier was a pure scientist—lacking in those qualities of aggressiveness and perseverance that would be required to win acceptance for the glacial theory. Resistance to the theory was strong; its defense must be equally so.

While the well-known scientists of the day clung steadfastly to the established ice-raft theory developed by Lyell and supported by the very words of the Bible, many Swiss people had long ago accepted the glacial theory. The irony of this situation struck de Charpentier forcefully when, on his way to present a paper on the glacial theory before the 1834 meeting of the society in Lucerne, he found some unexpected support.

> Traveling through the valley of Hasli and Lungern, I met on the Brunig road a woodcutter from Meiringen. We talked and walked together for a while. As I was examining a large boulder of Grimsel granite, lying next to the path, he said: "There are many stones of that kind around here, but they come from far away, from the Grimsel, because they consist of Geisberger [granite] and the mountains of this vicinity are not made of it."
>
> When I asked him how he thought that these stones had reached their location, he answered without hesitation: "The Grimsel glacier transported and deposited them on both sides of the valley, because that glacier extended in the past as far as the town of Bern, indeed water could not have deposited them at such an elevation above the valley bottom, without filling the lakes."
>
> This good old man would never have dreamed that I was carrying in my pocket a manuscript in favor of his hypothesis. He was greatly astonished when he saw how pleased I was by his geological explanation, and when I gave him some money to drink to the memory of the ancient Grimsel glacier and to the preservation of the Brunig boulders.

In spite of the woodcutter's toast, the theory was once again rejected by the society at Lucerne. But in the audience—and among those who rejected the theory—was Louis Agassiz.

Agassiz had first met de Charpentier while at school in Lausanne. Indeed, it may have been his great admiration for de Charpentier that influenced Agassiz in his decision to become a naturalist. Now, 10 years later, Agassiz himself was established as one of Europe's foremost men of science. Despite his admiration and liking for de Charpentier, however, Agassiz at first found it impossible to accept the glacial theory.

The older de Charpentier had often invited his young colleague to visit him at his home since the area around Bex boasted many fossils and geological features that he felt would be of interest to Agassiz. In 1836, two years after de Charpentier's presentation at Lucerne, Agassiz accepted de Charpentier's invitation and spent the summer at Bex. At this time, Agassiz was occupied chiefly with his research on fossil fishes. Although Agassiz, like the majority of scientists, subscribed to Lyell's ice-raft theory, he was not averse to seeing for himself whatever evidence his friend, de Charpentier, could show him in favor of the glacial theory. Agassiz went to Bex expecting to demonstrate to de Charpentier the fallacy of the glacial theory. Instead, he himself was speedily converted.

The gentle de Charpentier believed firmly in the idea that alpine glaciers had once extended far beyond their present limits, but he did not see it as the duty of a scientist to push for publication and widespread acceptance of this theory. De Charpentier was content to demonstrate the facts to friends and associates who visited him at Bex because he was confident that the theory would eventually prove itself. So it was that the glacial theory—born from the observations of simple peasants, developed by Ignace Venetz, and systematized by Jean de Charpentier—at last found a forceful spokesman in the person of Louis Agassiz (Figure 5).

Once converted, Agassiz was a quick and avid learner. In the company of de Charpentier and Venetz, he visited the glaciers of the Diablerets and the Chamonix valley, and the moraines of the Rhone valley. The evidence spoke for itself, and this time Agassiz listened. In a matter of weeks, he absorbed all that de Charpentier and Venetz could teach him. Agassiz soon outstripped his mentors. Using the facts that they had so painstakingly collected over the course of seven years, Agassiz quickly constructed a comprehensive glacial theory that he was confident would withstand the attacks of its enemies. Unfortunately, in his eagerness to present the theory, Agassiz took liberties with de Charpentier's work that the scrupulous gentleman found unforgivable. On several important points, Agassiz' expanded version of the theory went far beyond the available evidence.

In his excitement, Agassiz underestimated the opposition to the theory. The address that he presented to the Swiss Society of

Figure 5.   A portrait of Louis Agassiz at the Unteraar Glacier by Alfred Berthoud, now in the Library of the University of Neuchâtel. (From A.V. Carozzi, 1967, with permission of A.V. Carozzi and the University of Neuchâtel.)

Natural Sciences at Neuchâtel on July 24, 1837 had been hur-
riedly written the night before, and Agassiz was ill prepared for
the reaction he received. The members of the society had expected to hear news of their young president's research on fossil
fishes. They were somewhat startled when he launched into a
very different subject.

> Just recently, two of our colleagues [de Charpentier and
> Venetz] have generated through their investigations a con-
> troversy of far-reaching consequences for the present and
> the future. The characteristics of the place in which we meet
> today suggest my talking to you again of a subject which, in
> my opinion, may be solved by the investigation of the slopes
> of our Jura. I have in mind glaciers, moraines and erratic
> boulders.

Agassiz went on to describe in detail his own observations and
those of Venetz and de Charpentier. He interpreted these observations as evidence that masses of glacial ice once blanketed the
Jura. This ice, he said, was part of an immense polar ice sheet that
had covered Europe as far south as the Mediterranean, as well as
large parts of North America. Borrowing a term from his friend
Karl Schimper, a botanist, he described this period of the earth's
history as an ice age (*Eizeit*). The ice sheet was supposed to have
originated before the Alps were formed, then to have slid downward toward the Jura during a later uplift of the region. The
erratic boulders and polished rocks still visible in the area indicated the path taken by these moving masses of ice (Figure 6).

Agassiz' concept of an ice age shocked many in the audience. In
fact, Agassiz' "Discourse" caused such a furor that the scheduled
proceedings for the day were thrown into confusion. One timid
soul, Amanz Gressly, was so upset by the commotion that he never
got around to reading the manuscript he had brought with him
on the theory of sedimentation, later an important addition to
geology.

Agassiz' address succeeded in rousing strong emotions on both
sides of the glacier issue. In the lively discussion that took place in
the geology section afterwards, tempers ran high and sharp
words flew. Almost every scientist present found Agassiz' statements impossible to accept.

The meeting continued into the next day when Agassiz
brought up observations he had made in the Jura mountains

Figure 6. An illustration of polished bedrock near Neuchâtel, Switzerland published by Louis Agassiz in 1840. Agassiz argued that polished and grooved rock surfaces, occurring many miles from existing glaciers, were clear evidence of a former ice age. (From A.V. Carozzi, 1967, with permission of A.V. Carozzi and the University of Neuchâtel.)

around Neuchâtel itself. He also read an endorsement of the theory by Karl Schimper. But resistance to the theory was still strong. With the arrival of Elie de Beaumont, the opposition closed ranks.

Agassiz was sure that even the toughest of skeptics could not fail to be convinced, as he himself had been, by the evidence contained in the rocks themselves. A field trip to the Jura mountains was planned for the following day, and arrangements were hastily made for the members of the society to travel by carriage from Neuchâtel to La Chaux-de-Fonds in the heart of the Jura. An amused participant later wrote:

> In general, I was convinced by my short acquaintance with the leading scientists of the party that a great amount of jealousy and egoism existed between them. Elie de Beaumont was, during the entire trip, as cold as ice. Leopold von Buch was walking straight ahead, eyes on the ground, mumbling against an Englishman who was talking to Elie de Beaumont on the Pyrenees while we were in the Jura, and complaining rather offensively about the stupid remarks made by some amateurs who had joined the group. Agassiz, who was probably still bitter about the sharp criticisms made by von Buch of his glacial hypotheses, left the group immediately after departure and was walking a quarter of a league ahead all by himself . . . .

Angry that his colleagues seemed to be unimpressed by the evidence of glaciation all around them, Agassiz may well have reflected unhappily that the long journey into the mountains, over the rough road with the tired horses, had been to no purpose after all.

If that is what he thought, he was wrong. For his "Discourse," the field trip, and his monumental *Studies on Glaciers* (published in 1840), would at long last focus the attention of the scientific world on the issue of ancient glaciation. Despite some exaggerations, the courageous address Agassiz had delivered at Neuchâtel in 1837 served an important purpose. From that time on, however forcefully its opponents might argue against it, the glacial theory could no longer be ignored.

In the years that followed the Neuchâtel meeting, Agassiz continued his research on ancient glaciation, despite strong criticism from leading European scientists. In December 1837, Alexander von Humboldt urged Agassiz to return to his research on fossil fishes: "In so doing," he wrote, "you will render a greater service to positive geology, than by these general considerations (a little icy besides) on the revolutions of the primitive world, considerations which, as you well know, convince only those who give them birth."

It would be many years before von Humboldt came to realize that Agassiz—far from pursuing a chimera—had actually been among the first to recognize the truth. Now it would be Agassiz' task to convince other scientists that the earth had indeed experienced an ice age.

2

The Triumph of
the Glacial Theory

With his daring imagination, bold assertions, and vigorous prose
style, Agassiz had little difficulty capturing the attention of a wide
audience. Certainly statements like the following would attract
attention in any age:

> The development of these huge ice sheets must have led to
> the destruction of all organic life at the Earth's surface. The
> ground of Europe, previously covered with tropical vegeta-
> tion and inhabited by herds of great elephants, enormous
> hippopotami, and gigantic carnivora became suddenly
> buried under a vast expanse of ice covering plains, lakes,
> seas, and plateaus alike. The silence of death followed . . .
> springs dried up, streams ceased to flow, and sunrays rising
> over that frozen shore . . . were met only by the whistling of
> northern winds and the rumbling of the crevasses as they
> opened across the surface of that huge ocean of ice.

The idea that a catastrophe of awful intensity might once have
extinguished the life of the entire planet was not a new one. In
fact, it was commonly believed that the history of the earth was
divided into several epochs, each one of which had terminated in
a catastrophe powerful enough to deform existing layers of sediment and rock, set off floods of incredible magnitude, uplift
mountains, and destroy all plant and animal life on the planet. At
the beginning of each succeeding epoch, it was believed, new life
was breathed into the ravaged world—life which would survive
only until the next great cataclysm occurred.

Catastrophism was the dominant geological philosophy in the
eighteenth and nineteenth centuries because it neatly accounted
for the fossilized animal remains that were being unearthed by

geologists. The fact that the concept explained the fossil record without undermining the word of God, as set forth in the Old Testament, made it all but unshakeable.

It seemed obvious to scientists and laymen alike that the great flood that depopulated the earth, sparing only Noah and his ark full of animals, was in fact the catastrophe that had brought the last epoch to a watery close and launched the present one. For example, when the huge fossilized tooth of a mastodon was unearthed in a peat bog near Albany, New York in 1706, it was confidently judged to belong to one of the unfortunate and sinful people who had inhabited the earth before the Deluge of Noah. Governor Dudley of Massachusetts examined the specimen and then sent it off to the Boston preacher, Cotton Mather:

> I suppose all the surgeons in town have seen it, and I am perfectly of the opinion that it was a human tooth. I measured it, and as it stood upright it was six inches high lacking one eight, and round 13 inches, lacking one eight, and its weight in the scale was 2 pounds and four ounces, Troy weight . . . . I am perfectly of the opinion that the tooth will agree only to a human body, for whom the flood only could prepare a funeral; and without doubt he waded as long as he could keep his head above the water, but must at length be confounded with all other creatures and the new sediment after the flood gave him the depth we now find.

And 20 years later, in Switzerland, Johan Scheuchzer uncovered a collection of fossil bones in the deposits of an ancient lake. He concluded that they were the remains of an antediluvian man, destroyed by the Flood. Scheuchzer published a book under the title *Homo diluvii testis* ("The Man Who Witnessed the Flood"). Almost a century passed before the great French anatomist, Baron Georges Cuvier, examined the bones and correctly identified them as those of a giant, now extinct, salamander.

Agassiz himself had furthered the doctrine of catastrophism with his marvelously detailed illustrations of fossil fish and other extinct animals that were supposed to have lived during earlier epochs. Thus, in substituting an ice age for a flood, Agassiz challenged the established view of the nature of the last great catastrophe, but not the belief that a catastrophe had occurred.

A strong supporter of the flood theory—and an important ally for Agassiz if he could be won over—was the Reverend William Buckland of England. Since assuming the professorship of mineralogy and geology at Oxford in 1820, Buckland had become the most widely respected geologist in England. Like Agassiz, he had a flair for lecturing and created a stir wherever he spoke. Even at Oxford, a university noted for harboring eccentrics in many fields, Buckland was notable for the force of his personality and the oddity of his behavior. His classrooms were crammed to the rafters with a jumble of rocks, skulls, and skeletons that was famous throughout the university. Buckland believed in getting out of the classroom whenever possible and viewing geological deposits in their natural environments. On these excursions, he wore his academic robe and a dapper top hat—a habit that undoubtedly contributed to his popularity on campus. But, despite his eccentricities, Buckland was a dedicated and greatly respected scientist. Most of the leading geologists in England, including Charles Lyell, regarded themselves as his pupils.

Buckland was an ardent catastrophist. In his inaugural lecture at Oxford, "The Connexion of Geology with Religion Explained," he expressed his conviction that the objective of geology should be "to confirm the evidences of natural religion; and to show that the facts developed by it are consistent with the accounts of the creation and deluge recorded in the Mosaic writings." He was also the first scientist to devote most of his time to investigating the irregular accumulations of gravel, sand, clay, and large boulders that cover large areas of bedrock in Britain. Buckland's aim was to determine exactly how this seemingly chaotic deposit had been formed. Although there was no doubt in his mind that a flood had been the agent that had left the deposits, many questions remained to be answered.

Exactly how had the flood transported such vast quantities of debris? Buckland subscribed to the traditional view that floodwaters alone were sufficient to account for the diluvium (as it was termed by those who believed in this theory). In part, Buckland leaned towards this view because it accorded well with the biblical record. He was also convinced that this record was supported by evidence contained in the sediments themselves.

In 1821 a large number of strange bones had been discovered in a cavern in the Vale of Pickering. On hearing of this discovery, Buckland immediately traveled to Yorkshire to investigate. He found that most of the bones were those of hyenas, but scattered among them were bones of twenty-three other species including birds, lions, tigers, elephants, rhinoceroses, and hippopotamuses.

Buckland concluded that the cavern was an antediluvian hyena den that had been submerged in Noah's flood. He argued that the manner in which the bones had been covered with silt indicated that the animals had been drowned. From the quantity of post-diluvian stalagmites that had covered the floor of the cavern, he judged that the flood had occurred 5,000 or 6,000 years ago—a date, he noted, which was entirely consistent with genealogical records in the Bible.

Buckland presented his findings in a book dedicated to the Bishop of Durham: *Reliquiae Diluvianae; or, Observations on the Organic Remains Contained in Caves, Fissures, and Diluvial Gravel, and on Other Geological Phenomena, Attesting the Action of an Universal Deluge* (1823). Also included in this monumental work were the results of Buckland's studies of no fewer than twenty caves scattered over England and Europe. The book won for Buckland the Royal Society's Copley Medal and made him famous in geological circles.

Since none of these antediluvian caves contained human remains, Buckland concluded that the human species had been created only very recently. He was, therefore, somewhat shaken when the skeleton of a woman—dyed a rusty red and adorned with bits of ivory—was found in the deposits of a cave in Paviland on the southern coast of Wales. To many, this skeleton seemed to stand in direct contradiction to one of the primary tenets of the flood theory.

Buckland was able to explain the skeleton's existence, however, by pointing out that it had been encased in the uppermost layers of the sediment succession and that some clue to the reason for the lady's presence there might be found in the remains nearby of a Roman-age encampment. Putting two and two together, and suppressing a Victorian shudder of disapproval, Buckland wrote: "The circumstance of a British camp existing on the hill immediately above the cave, seems to throw much light on the character and date of the woman under consideration; and whatever may have been her occupation, the vicinity of a camp would afford a motive for residence, as well as the means of subsistence, in what is now so exposed and uninviting a solitude."

But other features of the diluvium were not as easily explained as the skeleton of a scarlet woman. Chief among these were the erratic boulders, many the size of small houses, which had been transported from their original locations hundreds of miles away (Figure 7). In addition, scratches and grooves displayed by the bedrock surface underlying these deposits, and the unsorted nature of the deposits themselves, were puzzling indeed.

Some catastrophists argued that such phenomena were the result of huge waves of a very special type that had never been observed. The dynamics of these "waves of translation" were elaborately analyzed by mathematicians at Cambridge who carefully calculated depths and velocities and published their conclusions in scholarly journals.

Other geologists did not believe that violent currents could have been the agents responsible for moving these huge erratics. They espoused a version of the flood theory proposed by Charles Lyell in the 1833 edition of his influential textbook *Principles of Geology*. Lyell suggested that the boulders might simply have become frozen in icebergs and slowly drifted to their present erratic locations. Devotees of this iceberg theory, which nicely preserved the idea of a universal flood, named the deposit "drift" to indicate the method by which it had been transported.

Additional support for the iceberg theory was found in reports by explorers in the north and south polar regions. No less a figure than Charles Darwin, in the *Journal* (1839) of his voyage on the *Beagle*, observed that some icebergs he had seen in the southern ocean contained boulders.

But Buckland was the first to admit that neither Lyell's iceberg-drift theory nor the classical diluvial theory could provide explanations for all of the evidence. For instance, a rise in sea level of more than 5000 feet would be necessary to account for some of the drift deposited in mountainous regions. Where would this water have come from? Where would it have gone? In their frantic efforts to answer such questions, some diluvialists let their imaginations run riot—untrammeled by awkward facts. Waters gushed from underground reservoirs and disappeared as suddenly into uncharted caverns. The earth—wobbling on its axis—created tidal waves that swept up and over the highest mountains. Or a great comet had once grazed the earth's surface, causing watery convulsions of a magnitude never witnessed by humans.

Although Lyell's theory did not eliminate the "sea level problem," it could be modified to explain some drift observed at high

Figure 7. Erratic boulder in Scotland. Louis Agassiz attributed the occurrence of large boulders, many miles from a possible bedrock source, to the action of ice-age glaciers. (From J. Geikie, 1894.)

elevations. To account for the erratic blocks in the Jura mountains, for example, Lyell invoked not icebergs floating on the ocean, but ice rafts drifting in large lakes—lakes that had been formed when rivers had become dammed by earthquakes or avalanches.

From Buckland's journals, it is clear that he was not completely satisfied with the answers provided by either the flood theory or the iceberg theory, and he continued his search for a means of explaining every aspect of the drift. Then, in September 1838, he attended a meeting of the Association of German Naturalists in Freiberg, Germany. There he listened as his friend, Louis Agassiz, presented forceful arguments in support of the ice-age theory first presented the year before at Neuchâtel. Buckland had heard rumors of Agassiz' radical theory and had come to Freiberg with the intention of examining the evidence at first hand.

After the meeting, Buckland and his wife traveled to Neuchâtel—to the mountains that had not long before convinced Agassiz himself of the truth of the ice-age theory. There were two other people in the group of travelers. Agassiz, excited at the prospect of converting the influential Buckland, was one. The other was Charles Lucien Bonaparte, Prince of Canino and brother of the former French emperor, Napoleon. Charles was a wealthy man with a passionate interest in natural history and little else with which to occupy his time since the French defeat at Waterloo in 1815.

Buckland had arranged the visit with Agassiz for personal as well as scientific reasons. He and his wife had met the Swiss naturalist several years earlier when they had offered Agassiz the hospitality of their home while he toured England studying collections of fossil fish. The three had become fast friends, and now the Bucklands looked forward happily to meeting Agassiz' young wife, Cecile, in Neuchâtel.

As he traveled through the mountains to Neuchâtel, Buckland's mind must have been busy. What evidence could his friend show him that would convince him of the validity of the ice-age theory? The small party wasted no time in setting out into the mountains around Neuchâtel, and Agassiz led the way, pointing out the evidences of glaciation that he was sure would tell their own story. But Buckland remained stubbornly unconvinced. Finally, Agassiz led the group into the Alps where he hoped that the actual sight of the glaciers in action would convince the professor. Buckland was convinced but only temporarily. When Mrs. Buckland wrote to thank Agassiz for his hospitality, she added: "But Dr. Buckland is as far as ever from agreeing with you." Apparently, Buckland had had second thoughts once he was out of Agassiz' commanding presence and far from the evidence in the Alpine rocks.

Agassiz was disappointed at this turn of events, for Professor Buckland was a widely respected scientist. Once converted, the Oxford geologist would be as important to the glacial theory as the Emperor Constantine had been to Christianity. In fact, although Agassiz could not know it at the time, he did not have long to wait. In the fall of 1840, the tide began to turn in his favor.

The critical event was a trip to England that Agassiz took in the summer of 1840, primarily to study fossil fish. In September, he attended the annual meeting of the British Association for the Advancement of Science in Glasgow. There he read a paper summarizing his glacial theory, emphasizing once more that: "At a certain epoch all of the north of Europe and also the north of Asia and America were covered by a mass of ice."

Predictably, the reaction of most of the audience was negative. The leader of the attack was one of the outstanding geologists in Britain, Charles Lyell. Buckland himself remained silent, for reasons that are not known. But his journals indicate that he had recently reexamined the evidence in favor of the glacial theory. Perhaps the seeds sown by Agassiz two years earlier had simply needed time to germinate, or perhaps conversion came to Buckland as it did to Saint Paul—in a blinding flash of light. In any case, soon after the meeting, Buckland invited Agassiz and another well-known geologist, Roderick Impey Murchison, to join him on a field trip to study drift in Scotland and northern England. It was this trip that finally convinced Buckland that the theory so staunchly defended by his friend Agassiz was correct. Overnight, Buckland became the first major British convert to the theory. (Murchison, however, remained unconvinced and, for the rest of his life, argued strongly in favor of iceberg drift.)

One of Buckland's first acts as a new convert was to read the scientific gospel to Charles Lyell. This he managed to accomplish in surprisingly good time, for on October 15 he wrote to Agassiz triumphantly: "Lyell has adopted your theory *in toto*!!! On my showing him a beautiful cluster of moraines within two miles of his father's house, he instantly accepted it, as solving a host of difficulties which have all his life embarrassed him."

The ice-age theory was an idea whose time had come at last. Lyell, the newest convert, lost no time in preparing a lecture entitled: "On the Geological Evidence of the Former Existence of Glaciers in Forfarshire," which he presented at the November meeting of the Geological Society of London. Agassiz himself presented a paper: "Glaciers and the Evidence of their having Once Existed in Scotland, Ireland, and England." And this time, Buckland came forward to defend the theory with his paper on the "Evidence of Glaciers in Scotland and the North of England."

With this trio of internationally famous geologists proselytizing in favor of the ice-age theory, it might be supposed that all opposition would crumble. Far from it—the general reaction of the assembled scientists was quite negative, and a heated debate took place after the lectures by Agassiz and Buckland. According to notes taken by one observer, Buckland concluded the debate:

> . . . amidst the cheers of the delighted assembly, who were by this time elevated by the hopes of soon getting some tea (it was a quarter to twelve P.M.), and excited by the critical acumen and antiquarian allusions . . . poured forth by the learned doctor, who . . . with a look and tone of triumph,

pronounced upon his opponents who dared to question the orthodoxy of the scratches and grooves and polished surfaces of the glacial mountains . . . the pains of eternal itch without the privilege of scratching.

In science as in religion, belief is often the strongest in a recent convert. Less than a month earlier, Buckland had sat on his hands at the Glasgow meeting while the Agassiz theory was vigorously attacked. Understandably, his abrupt about-face at the London meeting did not go unnoticed. A popular cartoon showed the Oxford professor, complete with robe and geological kit, standing upon a scratched and deeply grooved bedrock pavement (Figure 8). Two specimens lie at the professor's feet bearing these labels: "Scratched by a glacier thirty three thousand three hundred and thirty years before the creation," and, "Scratched by a cart wheel on Waterloo Bridge the day before yesterday."

Despite the ready wit of the popular press and the adverse reaction of the members of the Geological Society, it seemed for a time that all of British geology would soon be converted to the Agassiz theory. In the following year, 1841, colleague Edward Forbes wrote to Agassiz: "You have made all the geologists glacier-mad here, and they are turning Great Britain into an ice house. Some amusing and very absurd attempts at opposition to your views have been made by one or two pseudogeologists." Events proved this report by Forbes a bit optimistic. It was another 20 years before the majority of British geologists had accepted the ice-age theory.

Why did this theory, whose validity now seems self-evident, encounter so much resistance 100 years ago? In part, the slow acceptance of the theory may be attributed to a natural resistance to new ideas—particularly if those ideas run counter to long-held scientific principles or to religious convictions. The Agassiz theory challenged both, although religious conviction was probably less of a factor than scientific orthodoxy.

For one thing, geologists had indisputable evidence that the ocean had overwhelmed land areas, not once but many times in the past. Fossil fish and fossil shells preserved in sedimentary rocks on every continent were ample proof of this. Page after page of Lyell's textbook is devoted to explaining these marine inundations and to establishing their geographic extent. The idea that the drift itself was evidence of a particularly turbulent flood

Figure 8. The Reverend Professor Buckland, equipped as a "glacialist." A contemporary cartoon by Thomas Sopwith, showing the Oxford professor well equipped to study glaciers, and standing on a scratched bedrock surface. In the original cartoon, the specimens at his feet are labeled: "Scratched by a glacier thirty three thousand three hundred and thirty years before the creation," and "Scratched by a cart wheel on Waterloo Bridge the day before yesterday." The version reproduced here was published by Archibald Geikie, who removed the labels in deference to his friend. (From A. Geikie, 1875.)

was the natural extension of a general and familiar principle.

In fact, it was the almost complete absence of marine fossils in the drift that led many researchers to doubt its marine origin. If this absence had been complete, the glacial theory would probably have been accepted much earlier. Unfortunately, however, some drift deposits did contain marine fossils, and these "shelly drifts" were a thorn in the side of glacial theorists like Agassiz. Shelly drifts are not widespread; they occur near modern coastlines in New England, in Germany, and in several places in Scotland and northern England. But they do exist, and in the mid-1800s diluvialists studied these marine fossils carefully and pointed to them as further proof that the drift that encased them has been transported not by glaciers but by icebergs floating in floodwaters.

The deposits of shelly drift succeeded in confounding even the staunchest defenders of the glacial theory until, in 1865, a Scot named James Croll was able to explain them as the work of ice sheets moving over areas that are now covered by shallow seas. The moving ice had scraped shells and mud from the sea floor and subsequently deposited them in their present locations. According to Croll, the fossilized sea shells are simply erratic boulders in miniature, transported from their submarine homes by glacial ice.

Another factor that worked against acceptance of the Agassiz theory was the general ignorance of glaciers among geologists. If these geologists found it difficult to understand glaciers, they found it next to impossible to imagine ice sheets of the magnitude that Agassiz postulated. Not until 1852 did a scientific expedition clearly establish that the Greenland glaciers form a huge ice sheet. It was only later in the century that the true dimensions of the Antarctic Ice Sheet were established (Figure 9). Inevitably, as these polar explorations proceeded, and as geologists working in mountainous areas observed the action of valley glaciers, it became easier for the scientific community to accept the idea that Europe had once been submerged under an ice sheet similar to those now found in Greenland and Antarctica. Predictably, those geologists who lived in mountainous regions of Scotland, Scandinavia, and Switzerland found this idea easier to accept than did those who lived in lowland areas near the sea. To this latter group of geologists, the concept of a marine flood seemed the most reasonable explanation for the drift deposits.

Figure 9. Antarctic Ice Sheet. As knowledge of the polar regions advanced during the nineteenth century, geologists were able to draw comparisons between modern ice sheets and conditions during the ice ages. (From J. Geikie, 1894.)

Yet another factor that worked against the glacial theory was the extravagance of Agassiz' assertions. In his excitement, the Swiss naturalist persisted in claiming a much greater geographic extent for the ice sheets than the evidence supported. In 1837 he stated that the ice had extended as far south as the Mediterranean. The fact that no drift deposits were ever found in these southern areas made it all the easier for skeptical scientists to reject the rest of the arguments Agassiz presented.

As the years passed, Agassiz' assertions about the extent of the ice-age glaciers grew ever more extravagant. In 1865, while on an expedition to South America, he discovered evidence that the glaciers of the Andes had once extended far beyond their present positions. From this evidence Agassiz drew the conclusion that the ice sheets that had covered Europe and North America had extended into the continent of South America as well. No hard evidence of such an extension existed, so Agassiz succeeded only in raising the hackles of geologists. Lyell wrote: "Agassiz . . . has gone wild about glaciers . . . . The whole of the great [Amazon] valley, down to its mouth was filled by ice . . . .

[Yet] he does not pretend to have met with a single glaciated pebble or polished rock." Fortunately, however, enough evidence had already been found in Britain and on the mainland of Europe to ensure the acceptance of the glacial theory by all but the most dedicated diluvialists.

While the scientific battle raged in Europe, Agassiz himself departed for America. The trip was planned at the urging of Charles Lyell, who had recently visited the United States and wanted the Swiss naturalist to see the New World for himself. Lyell waved farewell to Agassiz from the dock at Liverpool in September 1846, confidently expecting to see him again before another year passed.

After a rough crossing, Agassiz' ship docked briefly at Halifax before sailing on to Boston. Agassiz hurried ashore, eager to find evidence to support his theory: "I sprang on shore and started at a brisk pace for the heights above the landing . . . . I was met by the familiar signs, the polished surfaces, the furrows and scratches, the line engravings of the glacier . . . and I became convinced . . . that here also this great agent had been at work."

Agassiz was welcomed to Boston by John Amory Lowell, who invited him to live in his comfortable home on Pemberton Square. Like others before him, Lowell soon fell under Agassiz' spell. As a successful owner of a textile mill and member of the Corporation of Harvard University, Lowell was in a good position to make sure that the great European naturalist made Massachusetts his permanent home. Early in the following year, a professorship was created for Agassiz at Harvard. Agassiz, who by this time was in some financial difficulty, accepted the offer gratefully. America was his home until his death in 1873.

Agassiz traveled widely in his adopted country, and he was delighted to find that news of his glacial theory had preceded him. In fact, the theory had already been accepted by many American scientists. As early as 1839, just two years after Agassiz' lecture at Neuchâtel, the American paleontologist Timothy Conrad published a brief paper noting that "M. Agassiz attributes the polished surfaces of the rocks in Switzerland to the agency of ice, and the diluvial scratches, as they have been termed, to sand and pebbles which moving bodies of ice carried in their restless course. In the same manner I would account for the polished surfaces of rocks in Western New York." Two years later, the

State Geologist of Massachusetts, Edward Hitchcock, delivered an address on the subject of Agassiz' theory before the newly formed Association of American Geologists.

By the mid-1860s, some 30 years after it was proposed, the glacial theory was firmly established on both sides of the Atlantic. Scattered opposition was voiced for many years, the last attack being a 1000-page treatise published by the English eccentric, Sir Henry Howorth, in 1905. But none of the opposition could stand against the evidence in favor of Agassiz' theory. The existence of an ice-age world was now taken for granted. Serious research on that world was about to begin.

3

Exploring
the Ice-Age World

Convinced now that an age of ice had occurred in the past, geologists were eager to learn more. Like detectives examining the scene of a crime, geological investigators searched for clues that would enable them to deduce what had happened thousands of years earlier. But unlike the type of detective story in which the search for evidence is usually confined to one room, or at most to an English country estate, the evidence needed to solve this geological mystery was scattered over the face of the planet. A great deal of sleuthing needed to be done.

From this point of view, Agassiz had chosen a fortunate moment in history to propose his theory. For in the prosperous years of Queen Victoria's reign, the wealth generated by the industrial revolution and the resources of a far-flung empire made possible the organization of geological expeditions to the farthest corners of the earth.

Victorian geologists had both theoretical and practical motives for searching so persistently for evidence about the ice-age world. There was, of course, a natural desire to fill in the pieces of the puzzle with which Agassiz had presented them, but economics provided an additional motive. In every civilized country, geological surveys were organized to assess the potential economic value of little-known regions. Nowhere is this better illustrated than in the United States where, in the years following the Civil War, the West was explored and mapped by geologists on horseback. To carry on this work, the U.S. Geological Survey was created by an act of Congress in 1879.

In order to gain a clear understanding of glacial action, geologists went into mountainous regions to study active and recently active glaciers. In this way, they were able to learn how the ancient glaciers had operated and thus to solve the mystery of how deposits of drift had been formed. They discovered that glacial ice was formed by the piling up of layer after layer of snow. When this pile grew thicker than about 100 feet, sheer weight converted the bottom layers to ice. This ice then flowed sluggishly downhill, picking up loose material in its path and breaking off and assimilating huge chunks of solid bedrock. Stones and boulders that were frozen into the lowest layer of the glacier acted like the teeth of a giant file—smoothing and polishing, and sometimes scratching the rock pavement over which the ice moved.

Another important discovery made during this period of intense exploration was the law that controls the size of glaciers and the speed with which they flow. This law can be expressed in terms of a "snow budget." For any given climate, a glacier will maintain a certain size. How large the glacier is depends on how much snow falls each year and how much evaporates and melts. If the climate changes, the glacier will grow or shrink until it achieves a new balance.

What earlier geologists had not understood was that although the downhill edge of an equilibrium glacier is fixed, the rest of the glacier is constantly flowing downhill. In its uphill portions, where snowfall exceeds melting, the flow is rapid and eroded material is not deposited. Over the downhill portion of the glacier, however, where melting exceeds snowfall, the flow is slower and the glacier constantly deposits material on the surface underneath the ice. This material, lodged firmly in place and strongly compacted by the weight of the overlying ice, is called a lodgement till.

When the climate warms, the glacial margin seeks a new equilibrium position. In the case of a valley glacier, the equilibrium position is farther uphill. In the case of an ice sheet, the equilibrium position is farther toward the center of the sheet. But the lower part of the glacier then becomes stagnant. It ceases to flow and gradually melts away. Some of the stones, sand, and other material contained within this part of the glacier are thus released directly from the ice. This layer, called ablation till, is superimposed on the lodgement till. The rest of the sediment is carried away and deposited, as outwash, by streams of meltwater flowing within the stagnant glacier and along its margin.

Geologists in Victorian times were able to determine the extent of glaciers during the ice age by locating the thickest deposits of till. These consist of both lodgement and ablation layers and are known as terminal moraines. It was also discovered that some of the sediments that had been labeled "drift" were, in fact, outwash deposits that had been carried by meltwater streams and deposited in front of the glacier.

It took some time for geologists to discover that similar streams of meltwater operated in much the same way *within* the glacier—filling crevasses, subsurface tunnels, and caverns with irregularly shaped deposits of outwash sediment. Small wonder then that Agassiz' friend, Reverend Buckland, had been confused by these deposits. The blanket of sediment that the glacier left behind when it finally retreated was a chaotic jumble of unstratified deposits (those that were transported by the ice and then dropped helter-skelter over the landscape) and stratified deposits (those carried away by water and sorted and deposited in neat layers).

With all of this new information about glacial action at their disposal, it was not long before geologists were able to chart the ice-age world and make a map that showed the extent of the great ice sheets. In North America, the terminal moraine was found to be a continuous ridge, up to 150 feet high, that extended from eastern Long Island to the state of Washington (Figure 10). North of this terminal moraine, the glacial deposits were found to consist mostly of till. South of the moraine was a flat landscape formed by a blanket of outwash deposits.

In addition to plotting the margins of the ice sheets, geologists found that they were able to determine the flow direction of the ice by recording the positions of scratches and grooves that had been incised in the bedrock by moving glaciers. Sample readings taken over a wide area were compiled to give a comprehensive picture of glacial flow. Another way of accomplishing the same objective was to trace erratic boulders to their bedrock source. Then, simply by looking at a map, geologists could see what path the glacier had taken.

All of these techniques were employed not only in North America, but also in Europe, Asia, South America, Australia, and New Zealand. By 1875 this effort had resulted in a global map that told the story of the great glaciers as they existed at the height of the ice age.

Worldwide, the glaciers had covered approximately 17 million square miles—three times the area covered by ice today. But, since the glaciers were confined almost entirely to the northern

I DEAL MAP
of
NORTH AMERICA
during the
ICE AGE

T.C.Chamberlin
1894

Figure 10. Professor T.C. Chamberlin's map was the first attempt to picture North America during the last ice age. (From J. Geikie, 1894.)

hemisphere, this figure is somewhat misleading. In the northern hemisphere alone, glaciers covered approximately 10 million square miles—almost 13 times the area they cover in that hemisphere today. But the southern hemisphere then was dominated—as it is now—by the 5-million-square-mile mass of the Antarctic Ice Sheet, which expanded only slightly during the ice age. The ice sheets that occurred elsewhere in the southern hemisphere were actually only small ice caps, and these spread out only slightly from mountainous areas in the southern Andes and from the mountains of southeast Australia, Tasmania, and southern New Zealand.

Victorian geologists were surprised to find that the great ice sheets in the northern hemisphere had a northern as well as a southern boundary. Thus, Agassiz' idea that a single great ice sheet had spread out to cover most of the northern hemisphere from a center at the North Pole was found to be incorrect. In fact, individual ice sheets had expanded from different spreading centers. The Laurentide Ice Sheet, for example, grew from a spreading center near what is now Hudson Bay (at a latitude of only 60° N). From this center, the ice flowed northward towards the shore of the Arctic Ocean. Then as now, that ocean was apparently covered only by a relatively thin layer of floating sea ice.

As early as 1841, it became apparent to some geologists that if Agassiz' theory was correct, an enormous volume of water must have been subtracted from the oceans to build up the ice sheets on land. In a remarkably acute essay published in that year, the Scottish geologist Charles Maclaren noted: "There is a question arising out of the theory, which he [Agassiz] has not touched upon. If we suppose the region from the 35th parallel to the north pole to be invested with a coat of ice thick enough to reach the summits of the Jura, that is, about 5000 French feet, or 1 English mile in height, it is evident that the abstraction of such a quantity of water from the ocean would materially affect its depth." Using the scanty evidence then available to him, Maclaren went on to calculate that "the abstraction of the water necessary to form the said coat of ice would depress the ocean about 800 feet."

At the time, this estimate was considered to be wild speculation, but by 1868 enough information was available so that an accurate estimate of sea level during the ice age could be made. To make such an estimate, it was only necessary to discover the average thickness of the ice sheets whose boundaries were so clearly designated on the map. Geologists accomplished this by determining which mountains bore evidence of glaciation during the ice age and which did not. If a mountain top had been covered with ice, then the ice itself must have been at least as thick as the elevation of that mountain. An even more accurate assessment could be obtained from mountains (such as Mount Monadnock, New Hampshire) that had been only partly submerged by the ice sheet—their rocky, unglaciated summits protruding through the ice to form islands of rock in a sea of ice. Today, the landscape changes abruptly part way up such mountains. Below the critical point, the mountainside is smooth and even; above it, the topography is rough and uneven. The thickness of the ice sheet could be determined simply by finding the altitude of this critical point above the surrounding countryside.

From such calculations, geologists were able to estimate that continental ice sheets in the northern hemisphere were about one mile thick. Using this figure, they went on to make a rough estimate of the volume of ice. The first to make this calculation was Charles Whittlesey, a geologist from Cleveland, Ohio. In 1868, he described his aim:

> to show that there must have been, during the glacial period, a material depression of the surface of the ocean . . . Accumulation of ice . . . can only occur on land by deposition, in the form of rain and snow, which becomes congealed. The ultimate source of this deposition is evaporation from the open surface of the sea. Inland lakes, rivers, swamps, and low lands furnish vapor to the clouds; but all fresh water basins receive their supply originally from the ocean . . . . If the water deposition on the surface of the land is not returned to the sea, it must be subtracted from the common reservoir.

Whittlesey calculated that "at the period of greatest cold, the depression of the ocean level should be, at least, three hundred and fifty or four hundred feet."

The drop in sea level postulated by Whittlesey (and confirmed by recent studies) was large enough to have brought about significant changes in the geography of the seashore. Whittlesey wrote: "As the waters retired, the configuration of all the continents would change; groups of islands, like the West Indies, would unite, forming a smaller number of islands, but of larger area; new points would appear above ocean level, and large shoals ... become dry land." Later research by both archaeologists and geologists would show that it was across just such a land bridge— newly won from the sea—that Stone Age hunters from Asia first journeyed into the North American continent.

Before long, geologists in Scotland and Scandinavia found abandoned sea cliffs and other shore-line features indicating that sea level during the ice age was indeed much lower than it is today. And, in some places, they also found evidence that sea level immediately following the retreat of the glaciers was higher than it is today. This high shoreline is especially apparent in Scandinavia where, in the center of what is today a mountainous region, marine shell deposits are found at altitudes higher than 1000 feet. The Scottish geologist, Thomas F. Jamieson, was the first to interpret these marine deposits correctly. In 1865, he wrote that:

> In Scandinavia and North America, as well as in Scotland, we have evidence of a depression of the land following close upon the presence of the great ice-covering; and, singular to say, the height to which marine fossils have been found in all these countries is very nearly the same. It has occurred to me that the enormous weight of ice thrown upon the land may have had something to do with this depression.

Jamieson went on to suggest why this depression would occur. He postulated that underneath the earth's outer, rigid crust was a layer of rocks "in a state of fusion," which would flow under pressure.

This bold and original speculation was supported years later by geophysical measurements. Just as Jamieson suggested, the upper portion of the earth's crust was shown by the measurements to be floating on fluid material. When a quantity of ice is placed on the earth's surface, the crust sinks down—exactly as the addition of passengers in a rowboat causes it to ride lower in the water.

The shorelines of glaciated regions, therefore, tell a curious story of marine inundations. During the ice age itself, worldwide lowering of the sea level caused shorelines to move downward by about 350 feet. Simultaneously, the weight of the ice sheets depressed the land surface underneath them. When the ice sheets melted, there was an *immediate* response—a rise in sea level—and a *gradual* response—a slow uplifting of the land surface. Thus, in New England, Scandinavia, and other glaciated areas, deglaciation was followed immediately by flooding. With the passage of time, however, the land surface rose to its original height—causing the sea level to appear to drop. In some areas of the world, the land is still reacting to the removal of the ice. Around the shores of Lake Superior, for example, the land is rising at the rate of 15 inches per century. But, away from the heavily glaciated areas, the shorelines tell a much more straightforward story, reflecting only the general rise and fall of sea level as water was subtracted from, or returned to, the ocean reservoir.

While some geologists confined their studies to areas that had actually been covered by ice sheets, others investigated land areas away from these regions. These geologists discovered that more than one million square miles of Europe, Asia, and North America had been blanketed during the ice age with a layer of fine, homogeneous, yellowish sediment. Borrowing an old term used by German farmers, they called this deposit "loess" (pronounced to rhyme with "bus"). In some areas, this layer of silt was found to reach thicknesses exceeding 10 feet. In other areas, it was found only in thin, discontinuous patches.

The attention of geologists had first been drawn to this peculiar deposit early in the nineteenth century, but its origin had remained a mystery. The fact that loess was composed of minute, uniform grains of silt suggested that it might have been deposited by moving water. But the horizontal layering that characterizes other water-laid deposits is not present in loess. Moreover, marine fossils are absent. It was not until 1870 that geologists found an adequate explanation for loess. The explanation came from a German geologist, Ferdinand von Richthofen, who published his theory and later defended it to a skeptical colleague:

> It is perfectly evident that no theory starting from the hypothesis of the deposition of loess by water can explain all or any single one of its properties. Neither the sea nor lakes nor rivers could deposit it in altitudes of 8000 feet on hillsides. Origin from water is perfectly unable to explain the lack of stratification, . . . the vertical cleavage, the promiscuous occurrence of grains of quartz, the angular shape of

these, . . . the imbedding of land shells, and the bones of terrestrial mammals.

There is but one great class of agencies which can be called on for explaining the covering of hundreds of thousands of square miles . . . with a perfectly homogeneous soil . . . . Whenever dust is carried away by *wind* from a dry place, and deposited on a spot which is covered by vegetation, it finds a resting place. If these depositions are repeated, the soil will continue to grow.

Von Richthofen's explanation of loess as a wind-blown deposit became universally accepted. Geologists were able to clarify their picture of the ice-age world, and a new piece of the ancient puzzle slipped into place. When melting occurred at the southern boundary of the ice sheet, great quantities of silt were deposited by outwash streams. Because the deposits were neither covered with snow, nor held in place by vegetation, they were easily blown away by the high winds that swirled in front of the ice sheet. Von Richthofen's ideas were confirmed by observations in Alaska, where glaciers melt rapidly during the summer months and the great quantities of silt deposited at their base dry up and are blown away to cover nearby grasslands with fertile loess.

The silt that the ancient glaciers drained from Canada in melt-water streams has proved to be a boon to American farmers in the Midwest. For that silt was blown southwards where it settled and eventually became the rich, easily cultivated, and well-drained soil of America's farm belt.

Geologists working in the American West found evidence that parts of Utah, Nevada, Arizona, and southern California were wetter during the ice age than they are today. In 1852, Captain Howard Stansbury (a topographical engineer who was investigating the flatlands around Utah's Great Salt Lake) wrote these observations in his diary:

Upon the slope of a ridge connected with this plain, thirteen distinct successive benches, or water-marks, were counted, which had evidently, at one time, been washed by the lake, and must have been the result of its action continued for some time at each level. The highest of these is now about two hundred feet above the valley . . . . If this supposition be correct, and all appearances conspire to support it, there must have been here at some former period a vast inland sea, extending for hundreds of miles; and the isolated mountains which now tower from the flats, forming its western and southwestern shores, were doubtless huge islands similar to those which now rise from the diminished waters of the lake.

Subsequent research confirmed Stansbury's inference. During the 1870s, Grove K. Gilbert of the U.S. Geological Survey showed that the Great Salt Lake is only a remnant of a former and far more extensive lake, which he named Lake Bonneville (Figure 11). During the ice age, this ancient lake was larger than any of America's Great Lakes are today, indicating that the climate in the western part of the United States was not only colder but also significantly wetter than it is today.

When the era of exploration began, there were already strong hints that the earth had been glaciated not once but several times. As early as 1847, Edouard Collomb reported two layers of till in the Vosges mountains of France. But these were separated only by stream deposits that could be interpreted either as a record of a

Figure 11.   Shorelines of ancient Lake Bonneville, Utah. Terraces along the base of the mountains near Wellsville, Utah, were formed at various times during the ice age along the shore of Lake Bonneville—a vast body of fresh water that is no longer in existence. After the ice sheets disappeared, the climate of this region changed from humid to arid, and the level of the lake fell. The briny waters of Great Salt Lake are the only surviving remnant of Lake Bonneville. (From G.K. Gilbert, 1890.)

short and minor retreat of the glacial terminus, or as evidence of a major and prolonged period of glacial recession. In the 1850s, similar evidence was found in Wales, Scotland, and Switzerland, but the conservative view—that the intertill beds represented minor climatic fluctuations during a single ice age—was generally preferred.

In 1863, Scottish geologist Archibald Geikie argued that plant fragments found between layers of Scottish tills were clear evidence that sustained intervals of warm climate intervened between different glacial ages (Figure 12). Finally, in 1873, Amos H. Worthen, Director of the Illinois Geological Survey, showed that a humus-rich soil had developed on one till layer before being buried by another. Since soils of this kind can only develop when the climate is warm enough to support abundant plant growth, this was strong support for the supposition that warm interglacial ages had occurred. Only a few years later, John S. Newberry and W. J. McGee clinched the argument by showing that in the American Midwest, two sheets of till were separated by the remains of a former forest.

By 1875 geologists had completed their initial survey of what the world of the last ice age was like. They had mapped its glaciers; measured its sea level; and determined which areas had been cold and wet, which cold and dry. They had also discovered that the ice age was not a unique event—that, in fact, there had been a succession of ice ages, each separated by warmer, interglacial ages similar to the present one. With all of this behind them, geologists were ready to turn their attention from facts to theories.

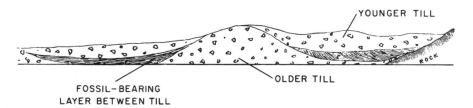

Figure 12. Multiple tills in Scotland. An exposure in the Cowden Burn railroad cut shows two layers of glacially deposited till, separated by a layer of fossil-bearing peat. Evidence of this kind was used by nineteenth-century geologists to prove the existence of more than one ice age. (From J. Geikie, 1894.)

PART II

EXPLAINING
THE ICE AGES

4

The Ice-Age Problem

Once they had accepted and enlarged upon Agassiz' glacial theory, geologists faced the challenge of explaining the ice ages. What agent stimulated the ice sheets to grow and expand? Why, having spread out to cover nearly one-third of the earth's land area, did those ice sheets then retreat? Most intriguing of all: would they return? These were the central questions of the ice-age mystery.

Many theories were advanced. Some that seemed at first to provide plausible answers were later rejected when fresh evidence proved them wrong. Others that were untestable had to be laid aside—judged according to the Scottish verdict, "Not proven."

Several attempts to solve the ice-age mystery ran into difficulty because they focused too narrowly on the fluctuations of the ice sheets themselves, failing to see them as only one part of a global climate system—a system that includes all of the mobile elements of the planet: ice sheets, ocean, and atmosphere. The three elements of this air-sea-ice system are interconnected in such a way that they behave like a huge machine. A change in one part brings about corresponding changes in the other parts of the system.

The energy that keeps the climate machine running—that causes the winds to blow and the currents to move—comes from the sun. Solar energy is received at every point in the atmosphere and at every point on the surface of the earth. Some of this energy is reflected back into space when it strikes dust particles or clouds in the atmosphere, or when it bounces off land or ocean surfaces. The rest of the energy received from the sun is absorbed and then radiated back into space. Every part of the climate system therefore gains a certain quantity of energy each day through absorption, and loses some through radiation and reflection.

At only two latitudes is there an exact balance between these gains and losses: 40° North and 40° South. At all other latitudes the radiation budget does not balance, and therefore, at these places, there is a tendency for the earth either to heat up or to cool down. Near the equator, the imbalance tends to raise the temperature. Land and sea surfaces there absorb much of the incoming radiation, days are long, and the sun is high in the sky. Near the poles, on the other hand, there is a net loss of heat because the ice and snow that are present there reflect much of the sun's energy. In addition, the sun's angle is low at these high latitudes. Unless processes other than reflection and radiation were at work, the poles would grow colder each year and the equator hotter. Winds and currents prevent this from happening by transporting heat from the equator toward the poles. Trade winds and hurricanes are examples of this heat-transport mechanism, as are the Gulf Stream in the Atlantic and the Kuroshio Current in the Pacific. Simultaneously, the south-flowing currents along the eastern sides of the North Atlantic and North Pacific transport cold water towards the equator.

Any valid theory of the ice ages must take into account that the growth or decay of a large ice sheet would have a large effect on the other elements of the climate system. For example, if an ice sheet is to expand, water must be drawn from the oceans, carried through the atmosphere to the site of the ice sheet, and precipitated there as snow. Variations in the volume of global ice are therefore linked inextricably with variations in sea level. Furthermore, any change in the area of an ice sheet must bring about a change in the radiation balance of the globe. When an ice sheet expands, heat is lost through reflection, global temperatures drop, and more ice is formed. Conversely, when an ice sheet shrinks, temperatures rise, and further shrinkage occurs. This "radiation-feedback effect" plays an important role in several theories of the ice ages because it explains how a small initial change in the size of the ice sheet is amplified.

The main objective of most theories is to discover the cause of this initial change. Since Agassiz' 1837 "Discourse," literally dozens of explanations for the ice age have been suggested. One of the earliest suggests that an ice age might be caused by a decrease in the amount of energy radiated by the sun. Because the climate system is powered by the sun, an ice age would indeed follow any significant decrease in radiation. However, there is no evidence at all that the energy output of the sun actually did decrease during any ice age. The main arguments in favor of the solar theory are, therefore, rather indirect. One approach points to observations made during the last century that show a slight tendency for the number of sunspots to be correlated with changes in rainfall and temperature. Unfortunately, it has never been proved that variations in sunspot numbers are actually correlated with variations in solar energy. Another argument, based on circumstantial evidence, suggests that over the past 1000 years, small advances of valley glaciers in mountain regions are correlated with changes in solar activity. Although small, these glacial fluctuations are associated with temperature changes on the order of 1° or 2° C.

But, even if the case for the solar control of climate were proved by observations made over the past century or millenium, such proof would not demonstrate that solar fluctuations are the cause of ice ages. Most investigators believe that the only way to test the solar theory would be to develop a way of calculating how the intensity of solar radiation varies through time. Until this is done, the idea that ice ages are caused by solar variations remains in limbo—neither proved nor disproved.

Another theory suggests that the uneven distribution of dust particles in space causes the climatic changes that set off an ice age. According to one version of this theory, when the earth passes through an area where dust particles are heavily concentrated, enough of the sun's energy is screened out to initiate a cooling trend. Another version holds just the opposite: when more dust particles fall into the sun, it glows more brightly than usual, causing temperatures on earth to rise. Clearly, these two versions of the theory must be reconciled before the dust-particle theory can be taken seriously. Supposing such a reconciliation were possible, there is still a major obstacle to organizing a definitive test of the theory. For, to date, astronomers have been unable to predict accurately how the concentration of dust between the earth and the sun has changed over the course of geologic history. If such a dust chronology were available, scientists could check it against the ice-age chronology. Then, if the two chronologies matched, the theory would be strongly supported.

The concentration of carbon dioxide in the atmsophere is the starting point for another ice-age theory. Although this gas occurs only in minute quantities (averaging about $^{33}/_{1000}$ of one percent), studies have shown that it exercises an important influence on global climate. This is because carbon dioxide has a peculiar property: while it is relatively transparent to the shortwave radiation that is received from the sun, it is relatively opaque to the longwave radiation that is reflected back into space. Changes in the amount of carbon dioxide in the atmosphere therefore force changes in the earth's heat budget. The more carbon dioxide there is in the atmosphere, the more the atmosphere behaves like the glass roof of a greenhouse—warming the environment inside by trapping incoming energy.

Many scientists are convinced that an ice age would result if the level of carbon dioxide concentration dropped low enough. But why should such a decrease happen? Until a theory is developed to explain why and how the concentration of carbon dioxide in the atmosphere might have altered over the course of the earth's history—and in particular, why this concentration should be lower when ice ages occur—the theory must be added to the list of plausible ideas that now seem impossible to test.

One of the more dramatic explanations of why ice ages occur supposes that they are initiated during epochs of frequent and explosive volcanic eruptions. During such epochs, the concentration of fine volcanic dust in the atmosphere increases and the dust reflects more of the sun's energy back into space, causing the earth's climate to cool.

Observations made following large volcanic eruptions have confirmed the validity of the basis of this theory. In 1883, for example, the East Indian volcano Krakatoa erupted with such violence that most of the island was destroyed and the noise of the explosion was heard 3000 miles away. Such quantities of dust were injected into the atmosphere that sunsets all over the world were noticeably redder for two years afterward. Careful measurements revealed that average global temperature dropped during this period as a result of the dust in the atmosphere. Eventually, the dust particles settled to the ground and climate returned to normal. Suppose, however, that the frequency of such volcanic explosions were to increase significantly. Might not the resulting cooling effect snowball into an ice age?

If volcanic dust was the agent responsible for causing the ice ages, evidences of it should be preserved in ancient soils, in existing glaciers, or in layers of mud formed in lakes and oceans at the time of these events. In principle, it should be possible to test the volcanic dust theory by comparing the historical record of ice-age climates with the sedimentary record of volcanic activity. In practice, however, it has so far proved to be impossible to collect measurements of sufficient accuracy and over a wide enough area to make a valid test.

According to another theory, developed by nineteenth-century English geologist Charles Lyell, ice ages are caused by vertical movements of the earth's crust. A general increase in land elevation would cause temperatures to drop, because the atmosphere is colder at higher altitudes. A more elaborate version of this early theory was presented in 1894 by an American geologist, James D. Dana. He envisioned not only a worldwide uplifting of land areas, but also the emergence of

> dry land, or a very shallow belt of water, across the North Atlantic from Scandinavia to Greenland [so that] the Arctic regions would have been deprived of a large supply of heat they now derive from the Gulf Stream. The confinement of the circuit of the Gulf Stream to the middle portions of the North Atlantic would concentrate thus its heat, make a much warmer ocean, and produce abundant precipitation.

But the Scottish geologist James Geikie (brother of Archibald) argued convincingly against this theory as early as 1874. He found it impossible to believe that such "vast oscillations of the crust could have taken place in so comparatively short a time." He added that, "if we find it hard to conceive of movements of elevation affecting simultaneously nearly the whole land surface of the northern hemisphere, our difficulties are not lessened by the reflection that we have still the glacial phenomena of the southern hemisphere to account for." All of the evidence compiled since Geikie's analysis has supported his negative reaction. There is today no reasonable basis for supposing that the theory is correct.

Another group of theories are of much more recent origin. These single out elements within the climate system of the earth to explain the ice ages. Among these theories, perhaps the best known was proposed by a New Zealand scientist, Alex T. Wilson, in 1964. He suggested that an abrupt sliding of large portions of the Antarctic Ice Sheet into the ocean would trigger the climatic changes that would lead to an ice age. In the normal course of events, the snow that accumulates on the surface of this ice sheet flows sluggishly outward towards its perimeter, where chunks of the ice break off and float away as icebergs. Wilson's idea was that the slowly increasing weight of the glacier and the accumulation of water along its base might, at intervals, cause it to collapse and to flow rapidly into the ocean. Sudden, periodic surges of this kind are known to occur in some mountain glaciers, and Wilson simply extended this idea to the Antarctic Ice Sheet. The climatic impact of a surge of this magnitude would indeed be very great. With each surge, the surrounding ocean would become covered with a highly reflective layer of floating ice. By reflecting back into space increased amounts of the sun's radiation, such a floating layer might be capable of triggering an ice age.

Many scientists were disappointed when no evidence was found to support Wilson's dramatic theory. If such a surge actually happened, sea level would rise dramatically. According to the theory, the ice age should have followed a sudden rise in sea level. But no evidence of such a rise has been found. Instead, the onset of the ice age saw a steady lowering of sea level (as explained in Chapter 3). Furthermore, a floating layer of ice like the one described above would leave a distinctive deposit on the sea floor when it melted. The fact that no such deposit has been found further discredits the theory.

Another theory based on variations within the earth's climate system was proposed by two scientists from Columbia University's Lamont Geological Observatory, Maurice Ewing and William Donn, in 1956. They argued that temperatures in the Arctic region are low enough today to permit the growth of an ice sheet—if snowfall were increased by a greater influx of moist air. If there were such an influx, the ice sheet would increase in size, and a self-perpetuating cooling trend would be initiated as the new snow cover increased the reflectivity of the earth's surface. But what agent would begin this trend by pumping moist air over Arctic areas that are now relatively dry?

The nub of the Ewing-Donn theory is the idea that an ice age begins when, for a brief period, the Arctic Ocean is free of ice and open to warm North Atlantic currents. During this interval, evaporation increases, the overlying atmosphere becomes charged with water vapor, and more snow falls on the surrounding lands. Once begun in this way, the reflection feedback effect escalates the process of glaciation, and leads to an ice age. The deglaciation process begins when temperatures drop low enough to cause the Arctic Ocean to freeze over again. With their main source of moisture cut off, the ice sheets shrink, sea level rises, and warm North Atlantic currents begin again to melt the Arctic sea ice.

According to this ingenious theory, the dynamics of the air-sea-ice system alone are capable of initiating a natural climatic cycle in which warm interglacial intervals alternate with glacial intervals. The theory is testable, for it predicts an historical sequence of events that would leave a record in the sediments of the Arctic Ocean. Layers of sediment that were formed in that ocean at the onset of the ice age should contain fossils of animals that had lived in sunlit waters. But no such fossils have been found. Instead, scrutiny of the sediments has shown that at no time over the past several million years has the Arctic Ocean been ice-free.

Other theories based on internal properties of the climate system are more difficult to analyze than either the surge theory of Wilson or the Arctic sea-ice theory of Ewing and Donn. Among these is the stochastic theory, which has found many supporters over the past decade. The central assumption of this theory is that large-scale variability is a natural and inherent property of climate.

On a short time scale, random variations in climate occur from month to month and from year to year. The stochastic ice age theory holds that the longer the interval examined, the greater the magnitude of these variations. Supporters of the theory point to the fact that climatic differences are found to be greater between succeeding decades than they are between succeeding years within the same decade. The theory assumes that, if still longer time intervals are examined, the magnitude of the observed variations increases without limit. Supporters of this theory offer sophisticated mathematical arguments to support this assertion.

According to the stochastic theory, any particular ice age does not require a special explanation. It is simply an example of a large-scale variation in the earth's climate, occurring as a cumulative effect of many small, random changes in weather stored in the oceans and in ice sheets. Because the stochastic theory holds that no particular event causes an ice age, it is difficult to test.

What's the score so far? Not very promising. Of the eight major theories proposed to explain the ice age, three were rejected, and the remaining five were untestable. But the game is not lost, for waiting to enter the lists is yet another theory, one which has been a serious contender ever since it was first proposed only five years after Agassiz' "Discourse" at Neuchâtel.

5

Birth of the
Astronomical Theory

The story of the astronomical theory begins with the publication in 1842 of a book called *Revolutions of the Sea*. This book was written by Joseph Alphonse Adhémar, a mathematician who made his living as a tutor in Paris. Adhémar was the first to suggest that the prime mover of the ice ages might be variations in the way the earth moves around the sun.

Adhémar knew that the earth's orbital path around the sun described not a circle but an ellipse, a fact which had been demonstrated in the seventeenth century by the astronomer, Johann Kepler (Figure 13). The axis of the earth's rotation is tilted 23½°

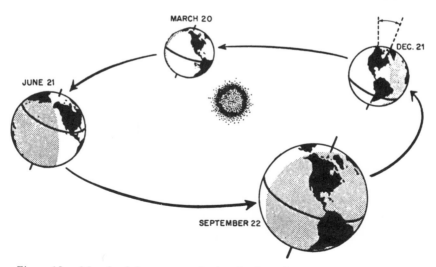

Figure 13. March of the seasons. As the tilted earth revolves around the sun, changes in the distribution of sunlight cause the succession of seasons. (Courtesy of G.J. Kukla.)

away from a vertical drawn to the plane of the orbit. Seasons occur because the orientation of the axis remains fixed in space as the earth revolves about the sun. When the North Pole points away from the sun, the northern hemisphere experiences winter. When the North Pole points toward the sun, that hemisphere experiences summer.

Kepler had shown that the sun is situated at one focus of the earth's orbit (Figure 14). The other focus is empty. As the earth travels around its orbit each year, it is sometimes nearer to, sometimes farther away from, the sun. Each year, on or about January 3, the sun reaches the point on its orbital path known as perihelion—the point at which it is closest to the sun. On or about

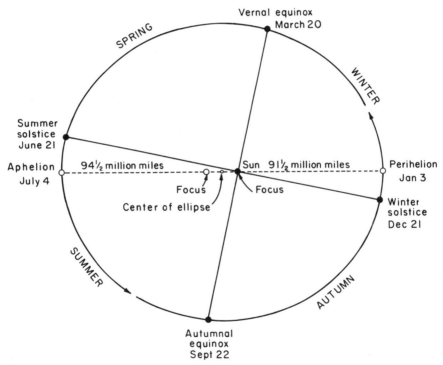

Figure 14. Dates of equinox and solstice. At the equinoxes, the earth's axis is pointed at right angles to the sun, and the day and night are of equal length all over the globe. At the summer solstice, the North Pole is tipped in the direction of the sun and the northern hemisphere has the longest day of the year. At the winter solstice, the North Pole is tipped away from the sun, and the northern hemisphere has the shortest day of the year.

July 4, it reaches aphelion—the point farthest from the sun. At aphelion, the distance between the earth and the sun is 3 million miles greater than it is at perihelion.

Each season begins at a particular point in the earth's orbit called a cardinal point. The earth reaches these points on or near December 21, March 20, June 21, and September 22 each year. In the northern hemisphere, December 21 marks the beginning of winter because the North Pole is tipped farthest away from the sun on that day, making it the shortest day of the year over the entire northern hemisphere. In the northern hemisphere, this point is called the winter solstice. In the southern hemisphere, December 21 is called the summer solstice because for that hemisphere it is the longest day of the year, and therefore marks the beginning of summer.

Six months later, on June 21, the earth reaches the cardinal point at which summer begins in the northern hemisphere and winter begins in the southern hemisphere. At this point, the North Pole is tipped toward the sun, making June 21 the longest day of the year in the northern hemisphere.

Only twice a year, on March 20 and September 22, are the two poles equidistant from the sun. On these dates, the number of daylight hours equals the number of hours of darkness at every point on the globe. These two cardinal points on the earth's orbit are therefore known as the equinoxes ("equal nights"). In the northern hemisphere, the vernal equinox (March 20) marks the beginning of spring, and the autumnal equinox (September 22) marks the beginning of fall. In the southern hemisphere, the seasons are reversed.

Two lines—one drawn through the equinoxes and the other through the solstices—would intersect at right angles to form a cross whose center is the sun. The short arm of the cross divides the orbit into two unequal parts. The distance traveled by the earth around one part of its orbit (from September 22 to March 20) is shorter than the distance traveled around the other part (from March 20 to September 22). In the northern hemisphere, therefore, spring and summer contain exactly seven more days than fall and winter. Thus, the total number of hours of daylight in the northern hemisphere exceeds the total number of hours of darkness by 168 hours each year (24 hours times 7 days). In the southern hemisphere, this situation is reversed. There, the warm seasons are seven days shorter than the cold seasons, and the number of hours of darkness exceeds the number of hours of daylight.

Adhémar argued that because the southern hemisphere has more hours of darkness each year than daylight, that hemisphere must be growing colder and colder—and he pointed to the Antarctic Ice Sheet as evidence that the southern hemisphere is in an ice age now.

Having explained to his satisfaction why the southern hemisphere is cold and partly glaciated today, Adhémar went on to try to explain why at other times in the past an ice age should have occurred in the northern hemisphere. He based his theory on the fact that over long periods of time, variations occur in the direction of the earth's axis. These variations were first discovered about 120 B.C. by Hipparchus, when he compared his own astronomical observations with those made by Timocharis, 150 years earlier.

Today, the point around which the stars seem to rotate (when viewed from the northern hemisphere) falls near the star Polaris—known as the Pole Star because the North Pole points toward it. Polaris forms the end of the handle of the Little Dipper. But in 2000 B.C., the North Pole pointed to a spot midway between the Little Dipper and the Big Dipper. In 4000 B.C., it pointed to the tip of the handle of the Big Dipper.

By plotting this progression on a star map, ancient astronomers were able to show that the North Pole does not always point in the same direction. Rather, the earth's axis of rotation wobbles, like that of a spinning top, so that the North Pole describes a circle in space (Figure 15). This movement is very slow—26,000 years must pass before the axis returns to the same point on the circle. In 1754, the French mathematician Jean le Rond d'Alembert explained that this phenomenon, called axial precession, occurs because of the gravitational pull that the sun and moon exert on the earth's equatorial bulge.

Axial precession causes the four cardinal points on the earth's orbit to move slowly around the orbital path. To an observer looking down on the North Pole, the direction of this movement would appear to be clockwise (Figure 16). Simultaneously, the elliptical orbit is itself rotating—independently and much more slowly—in a counterclockwise direction in the same plane. Together, the two movements cause the four cardinal points to shift along the orbit. This shifting is called the precession of the equinoxes.

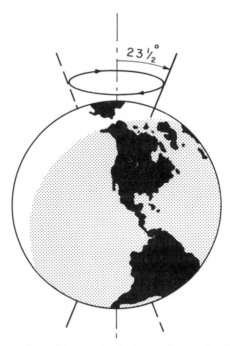

Figure 15. Precession of the earth. Owing to the gravitational pull of the sun and moon on the equatorial bulge of the earth, its axis of rotation moves slowly around a circular path and completes one revolution every 26,000 years. Independently of this cycle of axial precession, the tilt of the earth's axis (measured from the vertical) varies about 1½° on either side of its average angle of 23½°.

As shown by d'Alembert, the precession of the equinoxes completes a cycle every 22,000 years. Today, winter begins in the northern hemisphere when the earth is close to the sun near one end of the ellipse. Eleven thousand years ago, winter began when the earth was much farther away from the sun—near the opposite end of the ellipse. And 22,000 years ago, the earth was at the same orbital position it is today.

Adhémar theorized that glacial climates occur as a function of this 22,000-year cycle, and that whichever hemisphere had a longer winter would experience an ice age. Thus, every 11,000 years (every half cycle) an ice age would occur, alternately in one hemisphere and then in the other.

Most of Adhémar's theory was carefully reasoned. One inference, however, was so extravagant that it cast doubt on the theory as a whole. For Adhémar argued that the gravitational attraction

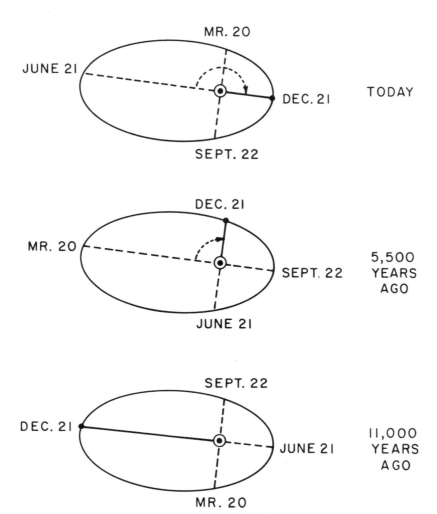

MR. 20

JUNE 21

DEC. 21

SEPT. 22

TODAY

DEC. 21

MR. 20

SEPT. 22

JUNE 21

5,500
YEARS
AGO

SEPT. 22

DEC. 21

JUNE 21

MR. 20

11,000
YEARS
AGO

● EARTH ON DEC. 21

⊙ SUN

Figure 16. Precession of the equinoxes. Owing to axial precession and to other astronomical movements, the positions of equinox (March 20 and September 22) and solstice (June 21 and December 21) shift slowly around the earth's elliptical orbit, and complete one full cycle about every 22,000 years. Eleven thousand years ago, the winter solstice occurred near one end of the orbit. Today, the winter solstice occurs near the opposite end of the orbit. As a result, the earth-sun distance, measured on December 21, changes.

of the Antarctic Ice Sheet was large enough to drain the water from the northern hemisphere ocean and create a sea-level bulge in the southern hemisphere. Painting a dramatic picture of what would happen when temperatures in the southern hemisphere began to rise, Adhémar predicted that the huge Antarctic ice cap would soften and rot until finally—eaten away at its base by the warming ocean—it would be left standing like a giant mushroom. At last, the entire mass would collapse into the sea, creating a huge iceberg-laden tidal wave that would sweep northward to engulf the land.

Although Adhémar's contemporaries dismissed these revolutions of the sea as mere fantasy, they did not find it so easy to criticize the astronomical part of the theory. The first to do so was Baron Alexander von Humbolt, a German naturalist, who in 1852 pointed out that Adhémar's basic idea—that one hemisphere is heating up while the other is cooling down—was incorrect. The average temperature of either hemisphere is controlled not by the number of hours of daylight and darkness, but by the total number of calories of solar energy received each year. And, as d'Alembert's calculations had demonstrated many years before, any decrease in solar heating that occurs during one season because the earth is farther from the sun, is exactly balanced by an increase during the opposite season, when the earth is closer to the sun. Therefore, the total amount of heat received by one hemisphere during the year is always the same as that received by the other.

The real reason for the greater coldness of the southern hemisphere was discovered many years later. Because the Antarctic continent is centered over the South Pole, isolated from other land masses, and far removed from the moderating influence of warm ocean currents, the continent is cold enough to support a permanent ice sheet. The ice sheet itself intensifies the cold temperatures by reflecting much of the sun's energy back into space.

Although Adhémar's theory was proved wrong, it was nevertheless an important step towards understanding the ice-age mystery. The idea that astronomical phenomena such as the precession of the equinoxes might have a significant effect on the earth's climate was not forgotten, and would set the stage for future discoveries.

6

The Astronomical Theory
of James Croll

When *Revolutions of the Sea* was published in 1842, the man who would eventually take Adhémar's ideas and develop them into a new astronomical theory of climate was working as a mechanic in the small Scottish town of Banchory. For James Croll, 21 years old and possessed of a deeply philosophical mind, the life was hard and unrewarding. In later years, he would remember that he "occupied, on the average, three different beds a week; and that these were not of the most comfortable character. We millwrights had generally to go to the ploughman's bothy [hut] . . . and frequently we had to bury ourselves under the clothes to secure protection from rats."

Croll's boyhood had been spent on the family farm in the tiny village of Wolfhill. His father, a stonemason, was away from home during the greater part of the year. When James reached the age of 13 he had to drop out of school and help his mother at home. Yet he managed to continue his studies by himself, and soon became engrossed in books on philosophy and theology. He later recalled his reaction to a book on physical science: "At first I became bewildered, but soon the beauty and simplicity of the conceptions filled me with delight and astonishment, and I began then in earnest to study the matter." Before long, his earnestness developed into an obsession to understand the basic principles of Nature.

> In order to understand a given law, I was generally obliged to make myself acquainted with the preceding law or condition on which it depended. I remember well that, before I could make headway in physical astronomy, I had to go back and study the laws of motion and the fundamental principles of

mechanics. In like manner I studied pneumatics, hydrostatics, light, heat, electricity, and magnetism.

By the time he was 16, Croll had obtained what he regarded as "a pretty tolerable knowledge of the general principles of these branches of physical science." To follow a career in science, however, he would have to obtain a university education; and this was far beyond his family's means. In the summer of 1837 he had to decide on an occupation.

> After several days' consideration, I thought I might try the occupation of a millwright. As I was fond of theoretical mechanics, it occurred to me that this occupation might be the most congenial . . . . But this I afterwards found to be a mistake; for, although I was familiar with theoretical mechanics, yet, as a working mechanic, I was scarcely up to the average. The strong natural tendency of my mind towards abstract thinking somehow unsuited me for the practical details of daily work.

This conflict between the practical demands of making a living and the desire to read and study would dominate Croll's life for many years. At last, in the fall of 1842, his scholarly yearnings won out. He resigned from his millwright job, and returned home to study algebra. The following spring he took up employment as a carpenter. Finding that this occupation suited him well, he decided to make carpentry his career. Unfortunately, an elbow injury that he had suffered as a boy, and which had never completely healed, became inflamed. By 1846, increasing pain forced Croll to seek a new occupation. Ready to turn his hand to anything, he worked for a time in a tea shop and eventually opened his own store. During this period, Croll met and married Isabelle MacDonald and the couple settled down to enjoy a comfortable life in Elgin.

As Croll later observed, "Strange are the ways of Providence, for had it not been for a mere accident in early life, I should in all likelihood have remained a working joiner to the end of my days." Instead, freed from the long hours of manual labor, he had time for reading and reflection. When he discovered a treatise by Jonathan Edwards on the philosolphical question of free will, Croll resolved "to commence at the beginning of the book and study it through, line by line, and page by page, until I should thoroughly master the treatise. This I did with the greatest care, often lingering for a day on a single page. It is probable that no one has ever devoted so much time to the study of the book as I have done."

Had Croll devoted as much energy to the shop as he did to his studies, his business might have prospered. His biographer and friend, James C. Irons, concluded that Croll was temperamentally unsuited to the role of shopkeeper. "To the day of his death," he wrote, "Croll had a modest, shy, dry, and almost speechless manner, except on occasions when he was drawn out by congenial conversation among real friends." Another friend agreed:

> It was something altogether extraordinary to see the man, with his large head, massive forehead, and kindly countenance, with his heavy form of body, hard horny hands and stiff arm, standing behind the counter of a tea-shop . . . . No one, even the most casual observer, could see Croll in the character of a shopkeeper at this time without knowing that he was not a shopkeeper to the manner born, and that he was evidently in a new sphere.

By 1850, his elbow joint had ossified completely and Croll was forced to sell his tea shop. For a while he made and sold electrical devices for the alleviation of bodily aches and pains, but the market for such devices quickly became surfeited, and in 1852 Croll turned hotelkeeper. To conserve his dwindling capital, he made most of the furniture for the hotel himself. The location Croll chose for his new venture was Blairgowrie—a town of only 3500 residents, situated far from any railroad, and already boasting a total of 16 inns and public houses. To make matters worse, Croll would not allow whiskey to be served in his establishment. Not surprisingly, the hotel failed and, in 1853, Croll found yet another occupation—this time as a life insurance salesman.

For the next four years, Croll sold insurance first in Scotland, and then in England. He later remembered this as the most disagreeable time of his life. "To one such as me, naturally so fond of retirement and even of solitude, it was painful to be constantly obliged to make up to strangers." Nevertheless, he persisted in this occupation until 1857, when his wife's illness forced him to resign from the Safety Life Assurance Company. The couple then moved to Glasgow where Isabelle could be nursed by her sisters. Unable for a time to find work, Croll "was now at perfect leisure," and "commenced to draw up some thoughts on the metaphysics of theism, a subject over which I had been pondering." He went to London and located a publisher for his manuscript. *The Philosophy of Theism* won favorable reviews, and both Croll and his publishers profited by it.

Two years later, Croll took a job as a janitor in the Andersonian College and Museum in Glasgow. "Taking it all in all," he later recalled, "I have never been in any place so congenial to me as that institution proved . . . . My salary was small, it is true, little more than sufficient to enable me to subsist; but this was compensated by advantages for me of another kind." For Croll now had access to a fine scientific library, and he was able to indulge his "almost irresistible propensity towards study, which prevented me from devoting my whole energy to business."

At first he concentrated on physics, and in 1861 published a scientific paper on electrical phenomena. Subsequently, however, his interest turned to geology. "At this period," he wrote later, "the question of the cause of the Glacial epoch was being discussed with interest among geologists. In the spring of 1864 I turned my attention to this subject." In the course of his studies, Croll came across Adhémar's book, published 25 years earlier. Although he realized that the French mathematician was wrong in believing that a change in the length of the warm and cold seasons could cause an ice age, Croll was convinced that some other astronomical mechanism must lie behind these geological phenomena.

Croll was familiar with recent studies by the great French astronomer Urbain Leverrier, which demonstrated that the degree of elongation of the orbit, known technically as orbital eccentricity, was slowly but continuously changing. Here was an astronomical factor that Adhémar had not considered. The French naturalist had built his ice-age theory around the precessional wobble of the earth's axis—and had assumed that the shape of the orbit itself remained unchanged. It occurred to Croll that changes in orbital eccentricity might be the real cause of the ice ages. Accordingly, he wrote a paper on the subject, which was published in the *Philosophical Magazine* of August 1864.

The paper excited a considerable amount of attention, and I was repeatedly advised to go more fully into the subject; and, as the path appeared to me a new and interesting one, I resolved to follow it out. But little did I suspect, at the time when I made this resolution, that it would become a path so entangled that fully twenty years would elapse before I could get out of it.

Croll's first step was to make himself familiar with the mathematical theory that enabled Leverrier to calculate how the eccentricity of the earth's orbit had changed. This theory was a direct application of Newton's law of gravitation. Each of the planets of the solar system exerts a force that tends to pull the earth out of its regular elliptical path around the sun. Because each planet revolves around the sun at a different speed, the combined gravitational pull the planets exert on the earth varies with time in a complex but predictable way. What Leverrier had done was to take the information then available on the orbits and masses of the planets, and calculate how the shape of the earth's orbit—as well as the degree of its axial tilt—had varied over the past 100,000 years. It had taken Leverrier 10 years to make the intricate calculations needed to pinpoint these variations on a time scale. Published in 1843, these calculations, based on the orbits and masses of the seven planets known at the time, led to the discovery of the planet Neptune.

Leverrier had measured orbital eccentricity by specifying the distance between the foci as a percentage of the long axis of the ellipse. As an ellipse approaches a circle, the two foci move closer together until the eccentricity of the ellipse becomes zero (Figure 17). Conversely, the more elongate the ellipse, the farther apart the foci become until the eccentricity reaches 100 percent. Right now, the earth's orbital path is only slightly eccentric (about one percent), Leverrier showed that the shape of the ellipse is constantly changing, so that in the past 100,000 years the degree of eccentricity has varied from a low near zero to a high of about six percent.

Using formulas developed by Leverrier, Croll calculated the orbital eccentricity for a sampling of dates over the past 3 million years, and drew a curve to illustrate the changes graphically (Figure 18). Croll was the first geologist to study the orbital history of the earth by plotting such curves. He discovered that the eccentricity of the orbit changes cyclically: intervals of high eccentricity, lasting many tens of thousands of years, alternate with long intervals of low eccentricity. Noticing that approxi-

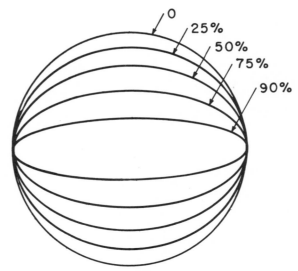

Figure 17.   Ellipses with different eccentricities.

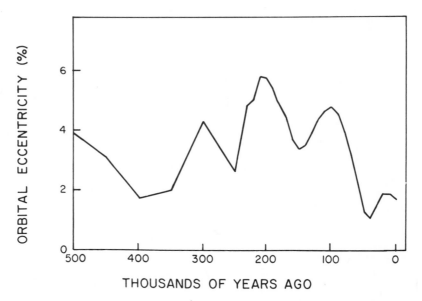

Figure 18.   Orbital eccentricities calculated by James Croll. According to Croll's theory, ice ages occurred during epochs of high orbital eccentricity. (Data from J. Croll, 1867.)

mately 100,000 years ago the earth's orbit was in a high-eccentricity state, and that for the past 10,000 years or so it has been in a low-eccentricity state, Croll concluded that there must be something about a highly elongate orbit that causes ice ages. He set out to discover what that could be.

His first efforts along this line were disappointing, for Leverrier had shown that the total amount of heat received by the earth during an entire *year* is virtually unaffected by variations in orbital eccentricity. Undaunted, Croll was able to demonstrate that the intensity of radiation received by the earth during each *season* is strongly affected by changes in eccentricity. He went on to construct a theory of the ice ages based on this seasonal effect.

Croll reasoned that a decrease in the amount of sunlight received during the *winter* favors the accumulation of snow. Furthermore, any small initial increase in the size of the area covered by snow must result in an additional loss of heat by reflecting more sunlight back into space. Therefore, Croll concluded, any astronomically induced change in solar radiation (however small) would be amplified by the snowfields themselves. Croll was the first scientist to develop this important idea, now referred to as "positive feedback."

Having proved to his satisfaction that winter was the critical season for producing an ice age, Croll went on to determine which astronomical factors control the amount of sunlight received during the winter. He concluded that the precession of the equinoxes must play a decisive role. If winter occurs when the earth is close to the sun (as it does in the northern hemisphere today), winters are warmer than usual. On the other hand, if winter occurs when the earth is far from the sun, temperatures are colder than usual (as they must have been in the northern hemisphere 11,000 years ago).

Although for a different reason, Croll had arrived at the same conclusion that Adhémar had reached 25 years earlier: every 11,000 years the precessional cycle results in a cooler winter climate in one hemisphere or the other. Croll then showed that changes in the shape of the orbit determine how effective the precessional wobble is in changing the intensity of the seasons. If the orbit were circular, for example, the precession of the equinoxes would have no effect at all on climate, for each season would then occur at the same distance from the sun. During such an epoch of zero eccentricity, winters would be of average intensity—neither exceptionally cold nor exceptionally warm. Croll noted that conditions today approximate this hypothetical case, for orbital eccentricity is only about one percent. This led him to conclude that during an epoch of low eccentricity, winters are not cold enough to induce an ice age no matter where the winter solstice occurs on the earth's orbital path. But during epochs of greater eccentricity, exceptionally warm winters result when that solstice occurs near the sun at the short end of the orbit; and exceptionally cold winters result when it occurs far from the sun at the long end of the orbital path.

Croll's theory takes into account both the precessional cycle and variations in the shape of the earth's orbit. It predicts that one hemisphere or the other will experience an ice age whenever two conditions occur simultaneously: a markedly elongate orbit, and a winter solstice that occurs far from the sun. Figure 19 shows how Croll thought these two factors worked together to change the earth's distance from the sun, and therefore its climate. When the earth is far away from the sun on December 21, an ice age occurs in the northern hemisphere. When it is close to the sun on that

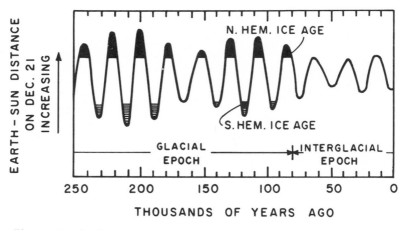

Figure 19. Croll's theory of the ice ages. Croll believed that ice ages are caused by changes in the distance between the earth and the sun, as measured on December 21. When this distance exceeds a critical value, northern hemisphere winters are cold enough to trigger an ice age; when this distance is less than a critical value, an ice age occurs in the southern hemisphere. During glacial epochs, orbital eccentricity is so large that these critical limits are often exceeded.

date, an ice age occurs in the southern hemisphere. The individual ice ages last about 10,000 years and occur first in one hemisphere and then in the other in response to the 22,000-year precessional rhythm. The long intervals during which the eccentricity of the orbit is great enough to cause ice ages in one hemisphere or the other are called Glacial Epochs; and the intervals that separate them are called Interglacial Epochs. According to this view, the last Glacial Epoch began about 250,000 years ago and ended about 80,000 years ago. Since then, the earth has been in an Interglacial Epoch.

Although Croll had no doubt that variations in the earth's orbit were responsible for changes in climate, he was worried that the magnitude of the climatic changes demonstrated by the geologic record might be too great to be explained by the rather subtle changes in orbital geometry—even if these were amplified by the reflection of sunlight. Was it really possible that an increase in orbital eccentricity of only two or three percent would result in the development of ice sheets massive enough to cover most of Europe and North America? Croll's concern anticipated objections that were later raised by other researchers. He attacked the problem with characteristic ingenuity by hypothesizing that the orbital changes operated as a triggering mechanism capable of setting off a major response within the earth's climate system. In the attempt to discover what this climatic response might be, Croll turned to the great warm currents of the Atlantic Ocean.

Today, the westward-flowing current that moves across the equator is deflected northward by the coast of Brazil and joins the Gulf Stream. In this way, heat is transported into the northern hemisphere from the southern hemisphere. But if some agency were to shift the Equatorial Current so that it struck the coast of Brazil south of its easternmost point, that warm current would be deflected southward, heat would be carried in the opposite direction, and the northern hemisphere would grow cooler.

What agency could be capable of shifting the Equatorial Current? To find an answer to this question, Croll developed an original (and essentially correct) theory that explained why the major ocean currents flow in the direction they do. Croll pointed out that the westward-flowing currents, which flow near the Equator, and the poleward-flowing currents such as the Gulf Stream, both move in response to the trade winds—much as the water in a teacup is moved by blowing on it—and that the velocity of these trade winds depends, in turn, on the temperature in the polar regions. If the polar region in one hemisphere were to become colder, more heat transport would be required to balance the earth's radiation budget in that hemisphere—and the trade winds there would blow harder. In short, the colder the poles, the stronger the winds. Croll concluded from this analysis that when the precessional cycle causes the ice sheets to expand in one hemisphere, the resulting increase in the strength of the trade winds in that hemisphere forces the warm equatorial currents in all of the oceans to shift towards the other hemisphere so that even more heat is lost. This effect, Croll felt, would be especially pronounced in low-latitude portions of the Atlantic Ocean, where a bulge in the coastline of Brazil would deflect the Equatorial Current either northward or southward. Thus, the direct climatic effect of any astronomically induced change in radiation— already amplified once by the reflection-feedback effect—is amplified once again by the changing pattern of ocean currents.

Croll could not have chosen a more propitious time to make his theory known. By 1864, William Buckland and Charles Lyell had been won over by Agassiz, the ice-age concept had been almost universally accepted, and geologists were eager to find an explanation for the ice-age cycle. Croll offered them a carefully reasoned theory that could be tested by comparing the geological record of climate with the astronomical predictions.

Among the many scientists who were impressed by Croll's paper was Sir Archibald Geikie, newly appointed Director of the Geological Survey of Scotland. Anxious to procure his services, Geikie urged Croll to leave his position in Glasgow and accept an appointment to the Geological Survey. In 1867, Croll accepted Geikie's offer, moved to Edinburgh, and continued his research.

In 1875, Croll published *Climate and Time*, a book that summarized his views on the cause of ice ages. In that book, Croll enlarged upon his original theory by taking into account Leverrier's calculation that the tilt of the earth's axis (as well as the eccentricity of its orbit) varies over the course of time. Now 23½° from the vertical, this value fluctuates within a range of about 3°, from a minimum of 22° to a maximum of 25°. Croll hypothesized that an ice age would be more likely to occur during periods when the axis is closer to vertical, for then the polar regions receive a smaller amount of heat. Unfortunately, Leverrier had not determined the timing of these variations in tilt, and Croll was therefore unable to follow up this important line of reasoning.

In the year after his book was published, Croll was made a Fellow of the Royal Society of London. Later, he received an LL.D. at the University of St. Andrews. The man who had begun his career as a mechanic in Banchory, operated a tea shop in Elgin, failed as an innkeeper in Blairgowrie, and worked as a janitor in Glasgow, had become a world-renowned figure in science (Figure 20). But fate would not smile on Croll for long. In

Figure 20.   Photograph of James Croll. (From J.C. Irons, 1896).

1880, at the age of 59, his brain was injured in what he referred to as a trivial accident, and he was forced to retire from the Geological Survey. From that time until his death ten years later, Croll fought an unsuccessful legal battle to win the full pension due him.

Eventually, thanks to several grants from scientific societies, the Crolls were able to move into a small house near Perth, where James continued to read and write in spite of almost continuous head pains. For five years he worked to improve his theory of the ice ages, but in 1885 he abandoned his scientific studies, and returned to his original passion—philosophy. In 1890 he published a small book, *The Philosophical Basis of Evolution*. In the same year, Croll was visited by friends who celebrated the occasion by taking a glass of whiskey. They were surprised to hear the abstemious scientist make "almost the only little joke" they ever heard him utter. "I'll take a wee drop o' that," he said, "I don't think there's much fear o' me learning to drink now." A few days later, aged 69, he died.

7

Debate Over
Croll's Theory

Croll's theory created an immediate and profound impression on the world of science. Here, at last, was a plausible theory of the ice ages that could be tested by comparing its predictions with the known geological record. Over the next 30 years Croll's ideas were widely and hotly debated: scientific expeditions were organized to dig for facts in drift deposits all over the world; articles in scientific journals probed the details of Croll's theory; and arguments pro and con filled many pages in geological textbooks.

James Geikie, professor at the University of Edinburgh in Scotland, and brother of Archibald Geikie, was among the first to come out strongly in support of Croll. His book, *The Great Ice Age*, published in 1874, was the first extended treatment of the ice-age problem since Agassiz' *Studies on Glaciers* in 1840. Geikie considered several competing theories, including the land-uplift idea of Lyell, and rejected them "because they all alike fail to meet the requirements of the geological evidence." He then went on to argue that the geological evidence strongly supported Croll's concept of an astronomically driven cycle of repeating ice ages.

Geikie based his argument on recently discovered evidence that the drift was not a simple deposit left by one glacier, as Agassiz had assumed, but a complex deposit composed of many separate layers of till—each layer representing the deposit of a separate glacial advance. In addition, the till sheets were often found to be separated by layers of peat containing seeds and leaves of plants that could not have survived in ice-age climates. Thus, the sedimentary succession left no room for argument—several ice ages had occurred in the past and these had been separated by warmer, nonglacial epochs. Just such a cycle of repeating ice ages had been predicted by Croll's theory.

In some places, only two tills were found, but in others it was possible to show that there had been at least six separate glacial ages, each followed by a warm interval. This evidence came primarily from Europe, but Geikie was careful to include in his book a chapter on North American glaciations written by the American geologist, Thomas C. Chamberlin. Chamberlin showed that North American drift deposits were composed of at least three layers of till—perhaps more. To emphasize this point, Geikie published a photograph showing three till layers of different colors, stacked one on top of the other at Stone Creek, Indiana.

At the same time that Geikie and his colleagues were unraveling the history of the ice ages as recorded in the unconsolidated deposits of glacial drift, other geologists were working to decipher the entire history of the earth as recorded in the rocks that lay beneath the drift (Figure 21). Between 1830 and 1865,

Figure 21. Succession of fossiliferous strata according to Charles Lyell. The entire history of the earth was subdivided by Lyell into named geological periods, each represented by a group of sedimentary rocks. The oldest period was called the Laurentian. (From C. Lyell, 1865.)

Charles Lyell had introduced a set of names to subdivide all of geologic time into Eras and Periods of unknown but presumably great length (Figure 22). With the discovery of the multiplicity of ice ages, and the recognition that the ice-age sequence must span a considerable interval of time, geologists naturally began to wonder where the ice-age cycle should be placed in Lyell's historical scheme. Presumably, the ice ages occurred during some part of Lyell's youngest Era, the Cenozoic. But how far back in time did they extend?

In 1846, Edward Forbes suggested that the post-Pliocene

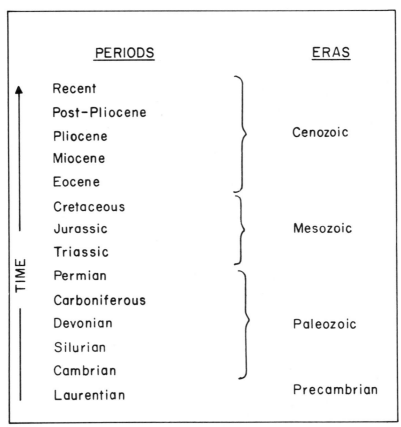

Figure 22. Lyell's classification of earth history. The geological periods shown in Figure 21 were grouped by Charles Lyell into eras. In an earlier version of his classification, Lyell had proposed the term "Pleistocene" for the period immediately following the Pliocene. By 1865, he had abandoned this practice. (From C. Lyell, 1865.)

Period must be the interval of time during which the glacial and interglacial deposits of the drift had been formed, and he recommended that the term Pleistocene—which Lyell had defined in quite another way only seven years before—be used as a substitute for the term post-Pliocene. Forbes's suggestion was widely adopted and forms the basis for modern usage of the term Pleistocene. Many geologists now also use the term Holocene Epoch (instead of Recent Period) to designate our present postglacial, or post-Pleistocene, time. In this usage, the Holocene Epoch and the

Pleistocene Epoch together constitute the Quaternary Period—
the Period during which there is a record of climate oscillating
between glacial and interglacial states (Figure 23).

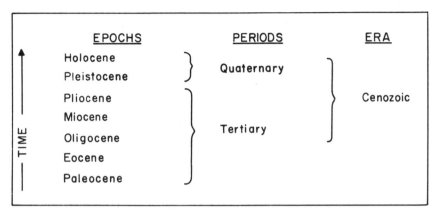

Figure 23.   Modern classification of the Cenozoic Era. In modern usage,
the term "Pleistocene" is used to designate a time interval immediately
following the Pliocene Epoch.

As the study of earth history progressed beyond the reconnaissance stage, geologists brought to light a new aspect of the ice-age
mystery. Evidence of glaciation was discovered in Paleozoic and
Precambrian rocks vastly older than Pleistocene drift. Geologists
now had to explain not only the relatively recent ice ages of the
Pleistocene Epoch, but the much older ice ages as well—and why
there were such long warm intervals between. An answer to this
question would not be found until well into the next century, as
described in the Epilogue. Meanwhile, the problem of testing
Croll's theory of the Pleistocene ice ages was challenge enough.

Although most geologists agreed with Geikie that the multiplicity of Pleistocene glaciations argued for the astronomical theory,
many others felt that the little that was known about the climatic
history of the southern hemisphere argued against it. The point
at issue was Croll's prediction that during periods of high eccentricity glacial ages would occur in alternate hemispheres every
11,000 years. If it could be shown that ice ages occur simultaneously in the two polar hemispheres, Croll's theory would be
disproved. Conversely, if it could be shown that glaciations did in

fact alternate between the hemispheres, the theory would be strongly supported.

Unfortunately, the problem of proving which of these historical scenarios was correct was beyond the capability of nineteenth-century science. The only sure way to correlate sedimentary layers in widely separated regions was to find at least one layer that could be traced continuously from one region to the other. Although ice-age deposits had been found in South America, Africa, and Australia, the oceans that separated these deposits from those in the northern hemisphere made correlation by tracing layers impossible. The only interhemispheric method that Geikie and his contemporaries could use was an unreliable one that was based on estimating how long a particular layer of drift had been weathered.

To test Croll's theory, the uppermost layers of drift in the two hemispheres were compared and many geologists concluded that there was little difference in age, thus disproving Croll's theory. But others pointed out that the amount of weathering is determined not only by age, but also by the availability of water, porosity of the sediment, average temperature, and other environmental factors whose impact is difficult to assess. Furthermore, they said, unless it could be shown that estimates of weathering intensity were accurate enough to distinguish between two tills with an age difference of only 11,000 years, the argument against Croll's theory could not be decisive. On this crucial point opinions were sharply divided. Geikie himself felt that the two hemispheres had been glaciated at different times. But he was cautious, and admitted that his opinion could not be proved. In contrast, Yale geologist James D. Dana ranged himself on the other side of the sticky question, concluding that "there is no evidence yet reported that the Glacial periods of the two hemispheres were not essentially simultaneous in their epochs."

The most powerful single test of Croll's theory would be a comparison of the dates he calculated for individual ice ages with actual dates determined from the geological record. Croll's theory would be strongly supported if it could be shown that the actual sequence of ice ages matched the sequence postulated by his theory. But how could the exact ages of glacial deposits be determined? Once again, the technical resources of nineteenth-century science were inadequate, and the problem of fixing an accurate chronology for the ice ages came to be recognized as the central problem in testing the astronomical theory. Only much later was the problem solved with sufficient accuracy to make a chronological test of the theory valid.

Meanwhile, nineteenth-century geologists made an ingenious attempt to estimate at least one date in the ice-age chronology by examining the landscape around Niagara Falls. As early as 1829, Robert Bakewell, Jr. had noted that the Niagara River flowed over a deposit of drift (a deposit that was naturally interpreted as the work of a great flood). He concluded that the river had occupied its present position only since the drift was deposited. Since that time, the flowing water had been eroding the rock layer, which forms the lip of the famous Falls, causing the position of the waterfall to recede slowly upstream and forming a spectacular gorge. Using observations made by long-time residents of the Niagara Falls area, Bakewell estimated that the rate of recession of the Falls is about three feet per year. By measuring the length of the gorge he judged that about 10,000 years had elapsed since the drift was formed. On his visit to Niagara Falls in 1841, Charles Lyell identified the drift as a glacial deposit, and then revised Bakewell's estimate of the recession rate to one foot per year. According to Lyell, therefore, the ice had begun to retreat about 30,000 years ago.

The matter was shelved until interest in Croll's theory focused the attention of geologists on the problem of measuring the length of postglacial time. Lyell then pointed out that both his estimate and that of Bakewell were a long way off the 80,000-year mark calculated by Croll. Several investigations were mounted with the aim of establishing a firmer date for the last glacial retreat, but the results were discouraging to those who believed in the astronomical theory. The revised estimates ranged from 6000 to 32,000 years, and turned many geologists against Croll's theory—despite warnings that the estimates were subject to large errors. American support for Croll's theory was further weakened when the state geologist of Minnesota, Newton H. Winchell, analyzed the recession rate of the Falls of St. Anthony on the Mississippi River near Minneapolis and concluded that the length of postglacial time was only 8,000 years.

By 1894, the majority of scientists in America were opposed to Croll's theory. The American view, undoubtedly influenced by the evidence gleaned from the falls at Niagara and at St. Anthony, was summarized by James D. Dana in his influential textbook: "[Croll's theory] is objected to by American geologists on the ground that the Glacial Period closed, according to American geological facts, not more than 10,000, or at the most 15,000, years ago, instead of the 150,000 or at the least 80,000 years ago, as the eccentricity hypothesis requires."

At the same time, the majority of European geologists followed the lead of James Geikie, who strongly favored Croll's theory. Geikie summarized his reasons as follows:

> The astronomical theory would appear to offer the best solution of the glacial puzzle. It accounts for all the leading facts, for the occurrence of alternating cold and warm epochs, and for the peculiar character of glacial and interglacial climates. It postulates no other distribution of land and sea than now obtains; it calls for no great earth movements the world over.

Strong support for these geological views came from the Irish astronomer Sir Robert Ball, who published a book in 1891 defending Croll's theory.

However, even Geikie had to acknowledge that the dates derived from the American waterfalls posed a serious problem. If these estimates proved to be correct—if the ice sheets really had disappeared from North America as recently as 6000 or even 10,000 years ago—Croll's theory would be seriously undermined. In that case, Geikie was prepared to defend the astronomical theory by retreating to the position that it applied only to the climate of Europe. For he was convinced that archaeological evidence in Europe and Asia proved that the ice sheets had retreated much longer ago than 6000 years. "No European geologist," he wrote, "will hazard the suggestion that the last great Baltic glacier existed at the dawn of civilisation in Egypt." Nevertheless, Geikie admitted that Croll's theory did not explain all of the geological facts, and he concluded his book with these prophetic words:

> The primary cause of these remarkable changes is thus an extremely perplexing question, and it must be confessed that a complete solution of the problem has not yet been found. Croll's theory has undoubtedly thrown a flood of light upon our difficulties, and it may be that some modification of his

views will eventually clear up the mystery. But for the present we must be content to work and wait."

As time went on, however, many geologists in Europe and America became more and more dissatisfied with Croll's theory, finding it at variance with new evidence that the last ice age had ended not 80,000, but 10,000 years ago. Moreover, theoretical arguments were advanced against the theory by meteorologists who calculated that the variations in solar heating described by Croll were too small to have any noticeable effect on climate. By the end of the nineteenth century, the tide of scientific opinion had turned against Croll, and his astronomical theory came to be treated as an historical curiosity, interesting but no longer valid. Eventually it was almost forgotten.

Almost, but not quite, for it would be revived years later by a Yugoslovian astronomer named Milutin Milankovitch. In 1890, when James Croll was on his deathbed in Scotland, Milankovitch was only 11 years old and unaware that he would be the one to pick up the thread of Croll's argument and interweave threads of his own devising to create an original design.

8

Through Distant Worlds and Times

Twenty-one years after the death of James Croll, and long after his orbital theory of the ice ages had been discarded, two young men—one a poet, the other an engineer—shared a table in a coffeehouse in Belgrade. They were celebrating the publication of a book of patriotic verse written by the young poet, and the small blue volume lay between them on the table. The poet's friend was Milutin Milankovitch, who recalled the occasion many years later in an autobiographical essay.

Coffee was all that the two friends could afford even for such a special celebration, but their spirits were high as they settled back in their chairs. They offered no objection when a well-dressed gentleman asked if he might sit with them, and when their new companion expressed a wish to glance through the book of verse, the poet acquiesced graciously. The gentleman proved to be a prominent bank director, as well as a fervent patriot, and he was so moved by the poet's verse that he ordered ten copies of the book and paid for them on the spot.

Now the two friends had something to celebrate indeed—and the means to do it in style. When the waiter approached with the steaming cups of coffee, they waved him away and ordered instead a bottle of red wine and a platter of cold cuts. When Milankovitch evoked the scene years later, he wrote that when "the first bottle was finished, the two friends were overcome by a feeling of joy. They had the sensation of being carried by invisible wings, and their field of vision seemed to grow ever wider. From the heights that they reached, they looked back on their activities and earlier achievements, which now seemed narrow and limited." By the time they had emptied the third bottle, the wine had "set their Southern blood in motion and filled them with confidence. With the self-assurance of an Alexander the Great, they sought out new areas to conquer; their Macedonia had grown too narrow for them."

The poet decided to stop writing short poems, resolving instead to devote himself to creating an epic novel. "In my new work," he said, "I want to describe our entire society, our country, and our soul." Not to be outdone, Milankovitch replied: "I feel attracted by infinity. I want to do more than you. I want to grasp the entire universe and spread light into its farthest corners." After sealing their resolutions with another bottle of wine, the two friends parted happily. The years to come would test the strength of their resolve.

Milankovitch had acquired his Ph.D. in 1904 at the Institute of Technology in Vienna. After graduating, he had been employed for five years as a practical engineer. He enjoyed this work, and found satisfaction in designing large and complex concrete structures, but he could not rid himself of a feeling that he should be working on more fundamental problems. When the University of Belgrade offered him a position as Professor of Applied Mathematics, he accepted. He was glad to be returning to his native Serbia (later part of Yugoslavia) despite the fact that his friends in Vienna thought him foolish to leave that sophisticated city to take up an academic post in provincial Belgrade. But Milankovitch knew that his country needed trained engineers—and he wanted a chance to tackle a more universal problem than the design of a concrete roof. He joined the faculty of the university in 1909, and taught courses in theoretical physics, mechanics, and astronomy. But he was still "under the spell of infinity and on the lookout for a cosmic problem" when, two years later, in the Belgrade coffeehouse, he made his decision.

Writing of this decision many years later, Milankovitch wondered if the wine might not have been responsible for his flash of inspiration. Whether or not the wine played a role, he had found the challenge he hungered for: he would develop a mathematical theory capable of describing the climates of earth, Mars, and Venus—today and in the past. Here was a problem large enough to absorb all of his talents and energy.

When Milankovitch discussed his new ambition at the university, he found that his colleagues were puzzled.

Our great geographer stared at me with an astonished look on his face when I told him of my intention to calculate the

temperature of the parallels of latitude on the Earth . . . .
Have we not built thousands of meteorological stations on
Earth that inform us more reliably and accurately about . . . .
temperature than the most perfect theory?

But to a theoretician like Milankovitch, the advantages of working with mathematical predictions rather than with thermometer
readings were obvious. For only theoretical calculations would
allow him to investigate temperatures in places beyond the reach
of direct observation: the upper atmosphere of the earth, and the
surfaces of the myriad moons and planets of the solar system.
"For the same oven, the sun, that supplies our Earth with heat also
heats those planets that are covered with solid crusts. Therefore
the results of the new theory would also apply to these planets. It
could give us the first reliable data about the climate of these
distant worlds."

And that was not all. For if it were possible to calculate the
climates of the planets as they are today, it would also be possible
to achieve the rest of his goal: to describe the climates of the past,
when the shape of the earth's orbit and the inclination of its axis of
rotation were different. In a word, the new theory "would make it
possible for us to cross the boundaries of our direct observations
in both time and space." But Milankovitch proceeded cautiously.
His first step was to discover what other researchers had done.

Milankovitch soon found that no one had accomplished what
he proposed to do. On the one hand, climatologists like his skeptical colleagues at the university had been content to make observations of temperature, rainfall, and wind velocity. On the other
hand, astronomers had confined themselves to determining what
the shapes of the planetary orbits were now, and what they had
been in the past, and had not attempted to calculate the distribution of radiation over the surfaces of the wobbling, tilting planets.
Although it was true that Adhémar and Croll—the pioneers of
the astronomical theory of the ice ages—had discussed the climatic effects of orbital variations at length, neither had had sufficient
mathematical training to calculate the magnitude of such effects
accurately.

Having determined that the path he proposed to follow had not
been traveled before, Milankovitch laid careful plans for his scientific journey into "distant worlds and times." Only a great
intellect could have conceived of such a venture. But it would take

more than a great intellect to carry it out: the journey Milankovitch proposed to make would take 30 years to complete.

Milankovitch worked on his theory for some part of every day. Even on holiday with his wife and young son, he carried several suitcases of books with him and insisted that a desk be provided in his room. In Belgrade, he carried out most of his studies at home, in a large study lined with books. (The room is preserved in its entirety by the Serbian Academy of Sciences.) On Tuesdays and Wednesdays he lectured at the university. Afterwards, he walked to his club where he met with his friends for an hour or so. At home, dinner was served at exactly eight o'clock every night, and was accompanied by a discussion of music or of world affairs. Two hours were spent at the table, another hour was occupied by reading. Finally, switching off the light, Milankovitch would sit in the dark and think.

Milankovitch laid the plans for his scientific attack with all the thoroughness of a general organizing an invasion. His first objective was to describe the geometry of each planet's orbit and to show how that geometry had evolved over past centuries. Milankovitch found, as Croll had before him, that three orbital properties determine how the sun's radiation is distributed over the planetary surfaces: the eccentricity of the orbit, the tilt of the axis of rotation, and the position of the equinoxes in their precessional cycle.

Milankovitch must have considered it a sign that he was destined to succeed when he discovered that only seven years earlier, in 1904, the astronomical calculations he needed had already been completed by the German mathematician Ludwig Pilgrim. While Croll had had at his disposal only Leverrier's calculations of variations in eccentricity and precession over the past 100,000 years, Milankovitch could make use of Pilgrim's calculations of variations in all three key properties (eccentricity, precession, and tilt) over the past 1 million years. Thus, the first important objective of Milankovitch's scientific campaign was easily achieved.

His second objective—to calculate how much solar radiation strikes the surface of each planet during each season and at each latitude—now seemed to be within easy reach. Two centuries earlier, Isaac Newton had worked out the general theory of radiation. Newton demonstrated that the sun's heating power depends on two geometrical factors: the distance from the sun, and the angle at which the sun's rays strike a particular part of a planet's surface. Because these geometrical factors could be derived from Pilgrim's results, Milankovitch concluded that it should also be possible to describe mathematically the distribution of solar radiation onto the surface of the planets.

Although simple in principle, the task of making these calculations proved to be enormously difficult in practice. For all of the planets are constantly spinning, revolving, wobbling and tilting in a crazy celestial dance, every movement of which has some effect on the radiation they receive from the sun. But Milankovitch was 32 years old and confident of his powers. Later he wrote: "I set out on this hunt in my best years. Had I been somewhat younger I would not have possessed the necessary knowledge and experience . . . . Had I been older I would not have had enough of that self-confidence that only youth can offer in the form of rashness."

At first, Milankovitch's investigations went well. "But in trying to penetrate more deeply into the problem, I ran into difficulties and could get no further," he wrote later. "Then [in 1912] the First Balkan War broke out. The Danube division of the Serbian Army, to whose staff I had been assigned, crossed the border of what was then the Turkish Empire early in the morning and was planning to capture Starac Mountain." As the young mathematician watched the Serbian troops fight their way toward the summit of the mountain, his thoughts turned to his own scientific attack, and the theoretical obstacles he had been unable to surmount. Then, as the Serbian regiment took the summit of Starac Mountain, the solution to his mathematical difficulties flashed into his mind, and he "conquered a mountaintop" on his own internal battlefield.

Two days later, the Turks were defeated. A cease-fire ensued, and Milankovitch was able to return to his private battle in the library in Belgrade. Although his progress now was rapid, he realized that it would be several years before his second major objective was attained. Keenly aware of the uncertain political atmosphere in the Balkans, he decided that he should go no further with his calculations until he had set down on paper the results he had obtained thus far. These were published as three short treatises during the years 1912 and 1913. Early in 1914, Milankovitch published another article, "On the Problem of the Astronomical Theory of the Ice Ages." Written in Serbian, and appearing in the midst of political turmoil in Europe, this article remained in scientific obscurity for many years. Yet Milan-kovitch's publications had already shed new light on the ice age problem by demonstrating mathematically that variations in orbital eccentricity and axial precession are large enough to cause ice sheets to expand and contract. Furthermore, he showed that the climatic effect of variations in the angle of tilt were even more important than Croll had suggested.

Satisfied that he had strengthened his rear guard, Milankovitch returned to his task confident that nothing could now stand in his way. All he needed was time to make his calculations. Then, in 1914, World War I broke out and Milankovitch was captured by the Austro-Hungarian Army while he was visiting his home town of Dalj. He was taken to the fortress at Esseg as a prisoner of war. Later, he recalled:

> The heavy iron door was closed behind me. The massive rusty lock gave a rumbling moan when the key was turned . . . . I adjusted to my new situation by switching off my brain and staring apathetically into the air. After a while I happened to glance at my suitcase . . . My brain began to function again. I jumped up, and opened the suitcase . . . . In it I had stored the papers on my cosmic problem . . . . I leafed through the writings . . . pulled my faithful fountain pen out of my pocket, and began to write and count . . . . As I looked around my room after midnight, I needed some time before I realized where 1 was. The little room seemed like the nightquarters on my trip through the universe.

On Christmas Eve 1914, the prisoner who had paroled himself by traveling through distant worlds received an unexpected but welcome gift—his freedom. His jailers had received a telegram from the Austro-Hungarian Ministry of War, ordering them to remove Milankovitch to Budapest. There he was released from custody on the condition that he report once each week to the police. A Professor Czuber, on learning that the talented Serbian mathematician had been imprisoned, had petitioned successfully for his release in the interest of science.

As soon as he was settled in Budapest, Milankovitch tucked his old leather briefcase under his arm, walked over to the library of the Hungarian Academy of Science, and knocked on the door. The director of the library, a fellow mathematician named Koloman von Szilly, received him with open arms. Milankovitch spent most of the following four years in the reading room of the library, working "without hurry, carefully planning each step." Two of those years were devoted to developing a mathematical theory for predicting the earth's climate as it is today. In the third and fourth years in Budapest, Milankovitch completed his description of the present climates of Mars and Venus.

Meanwhile, the war came to an end. Milankovitch gathered the labor of four years into his briefcase, boarded a white Danube steamer, and returned home to Belgrade. Despite the intervention of the war, he had attained his second objective—the mathematical description of the present climates on earth, Mars, and Venus. When these results were published in 1920, under the title *Mathematical Theory of Heat Phenomena Produced by Solar Radiation*, meteorologists soon recognized them as a major contribution to the study of modern climate. The book should also have been of interest to students of ancient climate, for it contained a mathematical demonstration that astronomical variations were sufficient to produce ice ages by changing the geographic and seasonal distribution of sunlight. In addition, Milankovitch claimed that it would be possible to calculate, for any time in the past, the amount of sunlight reaching the earth.

Although the book went unnoticed by the majority of geologists, it immediately attracted the attention of Wladimir Köppen, a widely known and greatly respected German climatologist. Köppen had compiled maps of the world that showed the geographical distribution of temperatures and precipitation, and had used this information to classify the earth's climate into zones that explained the geographic distribution of plant life.

Thus, the arrival of a postcard from the great Köppen caused no small stir in the Milankovitch household. Later, Milankovitch would write:

> One day, this simple postcard, which I saved like a relic, will be found in my estate. It came from Hamburg, from Wladimir Köppen, the great German climatologist, and referred to my recently published *Mathematical Theory*. Gradually it was followed by another forty-nine letters and cards, so that our mutual correspondence produced some one hundred writings. In his second letter, Köppen informed me

that, together with his son-in-law Alfred Wegener, he was working on a book about the climates of the geological past. Already seventy-six years old, this scholar had recognized earlier than others the benefits that could be derived for the paleo-climatological problem from my mathematical theory, and invited me to collaborate.

Milankovitch agreed willingly, and there followed a fruitful exchange of ideas between the Yugoslavian mathematician and the two Germans—one a famous climatologist, the other a leader among European geologists. Still a young man, Wegener had already made a name for himself with his theory that continents slowly shift their geographic positions. As Köppen had foreseen, Milankovitch's theory proved to be an invaluable tool in delving into past geological climates. But the collaboration brought benefit to Milankovitch as well, for he could hardly have found two people better equipped to guide him into the complexities of the geological record of climate.

Köppen was soon able to solve a major problem for Milankovitch. Having developed the mathematical machinery to permit calculation of solar radiation at any given latitude and season, Milankovitch was ready to attack his third major objective—a mathematical description of the earth's past climates. He planned to accomplish this by drawing a curve that would show the variations in radiation responsible for the succession of ice ages. But each circle of latitude—and each season at each latitude—had a unique radiation history. Milankovitch was therefore faced with the problem of determining which latitude and which season was critical to the growth of an ice sheet. Adhémar and Croll thought they had solved this problem by assuming that the critical factor was the radiation received at high latitudes during the winter. According to their view, ice ages occurred when the amount of radiation received by Arctic regions in winter was diminished. But Milankovitch was not convinced that this view was correct, and asked Köppen for his opinion. "After an exhaustive discussion of all the possibilities," Milankovitch wrote, "Köppen answered the question by indicating that it is the diminution of heat during the summer half-year which is the decisive factor in glaciation." He reasoned that changes in winter radiation could hardly have much effect on the annual snow budget, because temperatures in Arctic regions are cold enough for snow to accumulate even in modern times. During the summer, however, modern glaciers melt. Therefore, any decrease in the intensity of summer sunlight would inhibit melting, make the annual snow budget positive, and lead to glacial expansion.

Seeing the logic of Köppen's argument, Milankovitch began at once to calculate curves showing how summer radiation at latitudes 55°, 60°, and 65° North varied over the past 650,000 years. Even this task was no mean undertaking, for he writes: "I did my calculations for a full one hundred days from morning until night and then presented the results graphically by drawing three notched, curved lines to illustrate the changes in summertime radiation." He then mailed the graph (Figure 24) to Köppen, and eagerly awaited the climatologist's response.

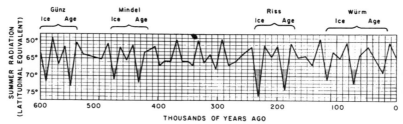

Figure 24. The Milankovitch radiation curve for latitude 65° North. The essential feature of the Milankovitch theory of the ice ages is a curve that shows how the intensity of summer sunlight varied over the past 600,000 years. In this version, originally published in 1924, Milankovitch identified certain low points on the curve with four European ice ages. Variations in radiation intensity are expressed in terms of latitudinal equivalents, for example, the radiation received 590,000 years ago at 65° N. is equivalent to that now received at 72° N. (Adapted from W. Köppen and A. Wegener, 1924.)

He did not have to wait long. Köppen immediately wrote that the pattern of notched lines on Milankovitch's graph could be matched reasonably well with the history of the Alpine glaciers as reconstructed 15 years earlier by the German geographers Albrecht Penck and Eduard Brückner. Köppen also expressed his wish to include the radiation curves in a book he was about to publish with his son-in-law, and invited Milankovitch to visit him at Innsbruck in Austria that fall to discuss the matter.

Understandably elated that the evidence supported his theory, Milankovitch agreed to meet his German friends in September

1924. Arriving at Innsbruck in time to attend a scientific meeting, he went immediately to the room where Alfred Wegener was speaking on "Climates of the Geological Past." The first part of the lecture dealt with Wegener's theory of continental drift and with the climates of remote geological periods. Later, Milankovitch would recall that Wegener spoke, "with the greatest modesty and in the simplest words . . . supported by an enormous number of facts." Not until Wegener began to talk about the climates of the Pleistocene Period,

> and pointed to my radiation curves that were projected on a screen did he raise his voice, because now he was discussing the work of another. He talked about my calculations with such excitement that I got quite embarrassed. I crouched in my seat in the top row of the amphitheater as much as I could so that a glance from Wegener would not reveal my presence in the auditorium.

That night, Milankovitch writes, he "slept on a bed of laurels and soft pillows." Milankovitch's stay in Innsbruck was not entirely devoted to work, however, for in the company of an old colleague from his engineering days, he managed to explore "all the pubs in Innsbruck."

The publication of Köppen and Wegener's book *Climates of the Geological Past* in 1924 assured a wide circulation for the radiation curves so laboriously computed by Milankovitch. Some geologists agreed with Köppen and Wegener that the curves fitted neatly with the geological record; others disagreed.

Milankovitch himself had no doubts on the subject. Returning to Belgrade after his memorable visit to Innsbruck, he plunged back into his studies. Thus far, he had produced curves only for the three latitudes 55°, 60°, and 65° North—portions of the globe thought to be the most sensitive to changes in the heat budget. Although changes in radiation would be less effective in lower latitudes, he reasoned that they would influence local climate to some extent. He therefore began to calculate radiation curves for each of eight latitudes ranging from 5° North to 75° North.

Milankovitch completed this work (the third major objective in his battle plan) in 1930 and published it as one volume of Köppen's *Manual of Climatology*. The title of the volume expressed Milankovitch's life-long goal so clearly that this time no geologist could miss its import: *Mathematical Climatology and the Astronomical Theory of Climatic Changes.*

With the publication of these eight radiation curves, geologists understood for the first time how two of the astronomical cycles influenced the pattern of incoming solar radiation. As Croll had foreseen, a decrease in axial tilt causes a decrease in summer radiation (Figure 25); and a decrease in the earth-sun distance at

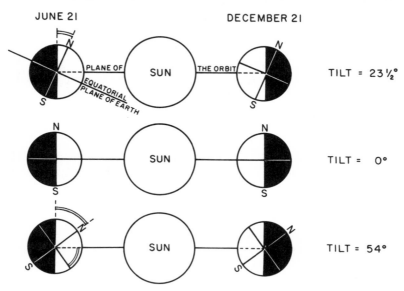

Figure 25. The effect of axial tilt on the distribution of sunlight. When the tilt is decreased from its present value of 23½°, the polar regions receive less sunlight than they do today. When the tilt is increased, polar regions receive more sunlight. The possible limits of these effects (never actually achieved) would be a tilt of 0°, when the poles would receive no sunlight; and 54°, when all points on the earth would receive the same amount of sunlight annually.

any season causes an increase in radiation at that season. But it was now clear that the strength of these effects varied systematically with latitude (Figure 26). The influence of the tilt cycle—the regular, 41,000-year oscillation of the inclination of the earth's axis—is large at the poles, and becomes small towards the equator. In contrast, the influence of the precession cycle—a 22,000-year oscillation of the earth-sun distance—is small at the poles and becomes large near the equator. Because the quantity

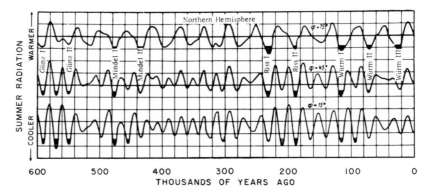

Figure 26. Milankovitch radiation curves for different latitudes. In 1938, Milankovitch published these curves showing changes in summertime radiation at 15°, 45°, and 75° North latitude. The effect of the 22,000-year precession cycle is clearly visible in the two low-latitude curves. Low points in the high-latitude curves are identified with the four named European ice ages. (Adapted from M. Milankovitch, 1941.)

of radiation received at any latitude and season is determined by the angle of tilt as well as the earth-sun distance, the shape of the radiation curve varies systematically from pole to equator. The curves calculated for high latitudes are dominated by the 41,000-year tilt cycle, while those for low latitudes are dominated by the 22,000-year precession cycle.

Milankovitch then began work on his fourth and final objective, which was to calculate how much the ice sheets would respond to a given change in solar radiation. Here the chief difficulty lay in estimating the importance of the reflection-feedback effect. That this mechanism operated to amplify any initial change in radiation had been known since it was first identified by James Croll. But so far all attempts at quantitative analysis had failed. Eventually, Milankovitch solved the problem by studying the altitude of the snowline—the elevation above which there is at least some snow all year long. Near the equator, the snowline occurs in the mountains, high above sea level. Near the poles, it occurs at sea level. Milankovitch was able to formulate a mathematical relationship between summer radiation and the altitude of the snowline, and thus to determine how much increase in snow cover would result from any given change in summer radiation.

In 1938, he published his results in a volume called *Astronomical Methods for Investigating Earth's Historical Climate*. Although the form of curves given in this volume did not differ greatly from the radiation curves published earlier, geologists now had a graph from which they could obtain an estimate of the latitude of the ice-sheet margin for any time in the past 650,000 years. Further, Milankovitch had made numerous additional calculations so that the jagged lines of his earlier graphs took on the form of smoothly changing curves.

With all four of his objectives accomplished, Milankovitch considered that his cosmic problem was solved. Ten years earlier he had begun writing a series of popular articles, which took the form of letters addressed to an anonymous young woman. Milankovitch had actually started this project many years earlier on a trip to Austria. The "letters" incorporated a great deal of autobiographical information, but their main purpose was to provide an informal introduction to astronomy and historical climatology. The early letters, published separately in literary magazines, achieved such popularity that in 1928 they were compiled and released under the title, *Through Distant Worlds and Times: Letters from a Wayfarer in the Universe*. The first edition was in Milankovitch's native language, Serbo-Croatian, but in 1936 the collection was expanded and published in German. The identity of the author's female correspondent remained a mystery— Milankovitch's wife asserting vigorously that she never existed at all.

During the late 1930s, Milankovitch began work on a comprehensive summary of his life's work to be called the *Canon of Insolation and the Ice Age Problem*. The final pages of this book were in the process of being printed on April 6, 1941—the day that Germany invaded Yugoslavia. In the upheaval that took place, the printing firm in Belgrade was destroyed, and the last pages of the book had to be reprinted.

The war itself did not worry Milankovitch unduly, for he was convinced that the Germans would lose. Moreover, he was filled with contentment and a quiet pride that the results of his long years of labor were now recognized internationally as a major achievement in science. True, there were some scientists who did not accept his theory, but he refused to defend himself in print, and later remarked that he did not regret this, "because without any help from me, several German scholars found the right answers to the objections that were raised." And he added that: "Today I have in my private library five independent scientific works and over one hundred treatises that use the radiation curves as a basis for their research into the course and the

Figure 27.   Milutin Milankovitch in a portrait painted by Paja Jovanovic
in 1943. (Courtesy of Vasko Milankovitch.)

chronology of the ice ages."

In 1941, at the age of 63, Milankovitch had completed his mathematical theory of radiation and had applied the theory to the ice-age problem. Years later, Vasko Milankovitch would remember his father telling him that:

> Once you catch a large fish, you cannot be bothered with the small ones any longer. For 25 years I was working on my theory of solar radiation and now that it is completed I am without work. I am too old to start a new theory, and theories of the magnitude of the one I have completed do not grow on trees.

One evening, over dinner, he announced to his wife and son: "I know what I will do during the German occupation. I am going to write the history of my life and my work. After my death somebody is going to write it and probably write it wrongly." He published these memoirs in 1952, and in 1957 completed work on a short synthesis of his scientific studies. The next year, aged 79, he died. The Yugoslavian mathematician, whose calculations had led him to wander in imagination through distant worlds and times, embarked on his final journey.

9

The Milankovitch
Controversy

With the publication of the Milankovitch theory in 1924, the attention of the scientific world was once again focused on the ice-age problem. Not since Agassiz first presented his glacial theory in 1837 had such widespread interest been expressed in the history of the earth; and not since the argument between Agassiz and Buckland had there been such an extended controversy over a climatic theory.

Fueling the controversy were the geological facts that had been gathered in the 60 years since James Geikie succeeded in settling the drift question. Geikie's work had inspired two generations of geologists to scour the globe for more evidence of past climate. Facts about the ice ages were abundant. What geologists needed now was a theory to integrate these facts and provide a comprehensive explanation of the ice ages. Milankovitch offered them just such a theory.

The most valuable feature of the Milankovitch theory was that it made testable predictions about the geological record of climate. It predicted how many ice-age deposits geologists would find, and it pinpointed when these deposits had been formed during the past 650,000 years.

These predictions were contained in three nearly identical radiation curves that showed past changes in summertime radiation at latitudes 55°, 60°, and 65° North (Figure 24). In theory, each radiation minimum caused an ice age. In all, there were nine minima, each appearing on the graph as a narrow projection extending well below the average level of radiation. Köppen and Wegener stressed the fact that these minima were not evenly spaced, but formed a distinctive, irregular pattern. The last three minima were grouped together, forming a triplet; these should correspond to ice ages 25,000, 72,000, and 115,000 years ago.

The other six minima were arranged in pairs. Milankovitch himself had pointed to the unusually long interval of high radiation that occurred about in the middle of the graph. He predicted that this interval would be represented in the geologic record by a very long interglacial age.

As soon as the astronomical theory was published, geologists familiar with the record of drift attempted to test the theory by counting the number of tills and by determining when they had been deposited. However, both objectives proved very difficult to achieve. Because each glacial advance tended to destroy the drifts of earlier glaciations, the succession of tills in most places was incomplete. Moreover, geologists had no accurate method of determining the age of any till. The best they could do was to make rough estimates of the duration of each ice age, and of each interglacial age, by noting the thickness and extent of the layers of till and soil.

Persevering in spite of these problems, geologists in North America—led by Thomas C. Chamberlin of the University of Chicago and Frank Leverett of the U.S. Geological Survey— concluded that there had been four major ice ages. The drift sheets corresponding to these ice ages were named for the states in which they were most easily studied. From the bottom up, the sequence of drifts was: Nebraskan (the oldest), Kansan, Illinoian, and Wisconsin (the youngest). Other geographic names were used to represent each of the interglacial intervals (Figure 28).

Armed with this impressive array of facts and names, geologists ranged themselves for or against the Milankovitch theory. Those who were pro-Milankovitch pointed out that the four North American drifts matched the four radiation groups (one triplet and three doublets) postulated by the theory. Opponents of the astronomical theory replied that since the ages of the North American drifts were known only within very broad limits, there was no way of being sure that any radiation minimum actually coincided with an ice age. The attempt to test the astronomical theory in this way was therefore inconclusive.

An entirely different approach to the problem of deciphering the ice-age succession had been developed in the 1880s by Albrecht Penck, a German geographer studying river valleys along the north slope of the Alps. Penck had discovered that the lower portion of each of these valleys was a flat surface (called a strath) that the river was eroding into a layer of gravel. At higher eleva-

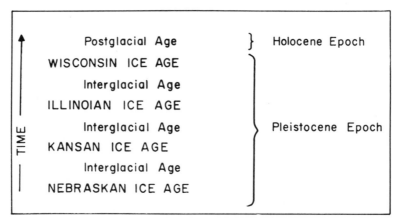

Figure 28. Theoretical succession of North American ice ages. By the end of the nineteenth century, glacial drifts corresponding to four Pleistocene ice ages had been recognized and named. Subsequent work has shown the existence of many more ice ages.

tions along both sides of the valleys Penck found three terraces— flat-topped embankments separated from each other by steep scarps. The terraces were composed of gravel layers similar to those which he had found on the valley floor. Penck had reasoned that each gravel layer had been formed during a cold climate when frost action and lack of vegetation had increased the rate of erosion. During intervals of warm climate, the rivers had evidently stopped depositing gravel, meandered from side to side, and cut a flat strath. Penck had concluded that the flat parts of the terraces were portions of former straths that had been formed during earlier interglacial intervals: the higher the terrace, the older the age of the corresponding interglacial. Each gravel deposit was therefore interpreted as a remnant of a more extensive layer that had been deposited during an ice age.

According to Penck's scheme, the sequence of Alpine gravel layers offered scientists what the drift sheets had not—a complete record of the glacial succession. Since there were four layers of gravel, there must have been four Pleistocene ice ages. And many geologists assumed, without proof, that the four American ice ages were the trans-Atlantic equivalent of the European succession.

In Europe, each ice age was named for a river valley. As a convenience to geologists, who were already overburdened with historical terms, the names were alphabetized: Günz, Mindel, Riss, and Würm. The oldest glaciation (the Günz) was represented by the gravels that composed the highest terrace. The most recent glaciation (Würm) was recorded by the gravels underlying the present river. Günz, Mindel, Riss, Würm—these names, coined by Penck and his colleague Eduard Brückner, would be stamped in the memories of generations of students, and would echo in lecture halls for years to come.

In addition to naming the succession of ice ages, Penck and Brückner were also able to estimate the length of time since the last ice sheet had disappeared from Switzerland. This they accomplished by studying the thickness of postglacial sediments in Swiss lakes, and by estimating how fast these sediments had accumulated. In this way the duration of postglacial time was calculated to be about 20,000 years.

With the 20,000-year estimate of postglacial time as a basis, Penck and Brückner proceeded to estimate the duration of earlier interglacials by comparing the depth of postglacial erosion with the depth of erosion that had occurred during each of the earlier warm periods. In this way they calculated that the interglacial that occurred immediately before the last (Würm) ice age was about 60,000 years long; and that the preceding interglacial—which they called the Great Interglacial—had lasted some 240,000 years. Altogether, they estimated that the Pleistocene was 650,000 years long.

In 1909 Penck and Brückner had published a curve that showed the history of Pleistocene climate (Figure 29). Fifteen years later, when Köppen received Milankovitch's radiation curves in the mail, he realized immediately that he could test the astronomical theory by comparing the radiation curves with

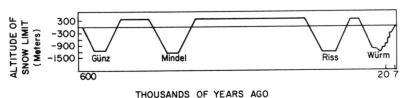

Figure 29. Theoretical succession of European ice ages, according to the climatic history of Europe suggested by the work of A. Penck and E. Brückner in 1909. During four supposed ice ages (named Günz, Mindel, Riss, and Würm), they estimated that the snow extended more than a thousand meters below its present level in the Alps. (Adapted from M. Milankovitch, 1941.)

Penck and Brückner's scheme. As noted in Chapter 8, Köppen and Wegener made this comparison in 1924 and concluded that theory matched fact amazingly well. In both the radiation diagrams made by Milankovitch and the climatic diagram drawn by Penck and Brückner, ice ages appeared as short pulses separated by longer warm intervals. Although the timing of the ice ages and the radiation minima did not agree in detail, the general pattern of the two curves was quite similar. Köppen and Wegener were especially impressed by the fact that the interglacial interval between the Mindel and Riss glaciations (the Great Interglacial of Penck and Brückner) was analogous to the long, warm interval predicted by Milankovitch. Finally, the 20,000-year date given by Penck and Brückner for the end of the last ice age matched reasonably well with the date of the last radiation minimum, 25,000 years ago.

Satisfied that Milankovitch's astronomical theory had been confirmed by an independent line of research, Köppen passed the good news along to Milankovitch and then published his curves in 1924. During the next fifteen years, the German geologists Barthel Eberl and Wolfgang Soergel restudied the Swiss terraces and discovered that several terraces recognized by Penck and Brückner were actually compound structures made up of more than one gravel deposit. But the revised version of Penck and Brückner's climatic curve seemed to match the details of the radiation curve even better than before; and Milankovitch included a summary of the geological work in his 1941 publication (Figure 30, next page).

During the 1930s and 1940s, most European geologists were won over to the Milankovitch theory. In fact, as Milankovitch himself noted with obvious pleasure: "A constantly increasing number of scientists undertook to classify the . . . sediments according to the new method, to connect them with the radiation curves and to date them by means of the latter." Subtly but surely, the emphasis had shifted: where once the geological record had been used to test the theory, now the theory was used to explain the record. "In this manner," Milankovitch said, "the ice age was given a calendar." Among those who developed this calendar was Frederick E. Zeuner, Professor of Geochronology at the University of London. In 1946 and again in 1959 he published books in which the Milankovitch calendar was used to date the main events of the Pleistocene Epoch.

Geologists in America, to whom the Alpine terraces seemed

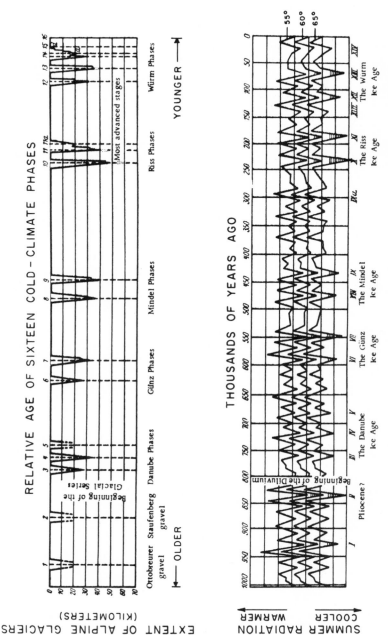

Figure 30. Eberl's test of the Milankovitch theory. Climatic history of Europe, as interpreted by B. Eberl (upper diagram) compared with the Milankovitch radiation curves for 55°, 60°, and 65° North latitude (lower diagram). Although the time scale for Eberl's climatic curve was rather uncertain, Milankovitch regarded the degree of correspondence between the two diagrams as proof of his theory of the ice ages. (Adapted from M. Milankovitch, 1941.)

remote and somewhat puzzling, were more skeptical. Even in Europe the theory was not unanimously endorsed. One man who spoke out against it was the German geologist, Ingo Schaefer. After devoting much attention to the Alpine river terraces, Schaefer became convinced that the fundamental hypothesis of the Penck-Brückner scheme was faulty. For he discovered that some layers of gravel contained fossil molluscs that are found today only in warm climates. How could sediments containing such fossils have been deposited during an ice age? It was a sticky question, threatening to undermine the very foundations of the Milankovitch theory. Most European geologists chose to ignore the problem, dismissing Schaefer's fossils as insignificant exceptions to a general rule.

Before long, however, other voices were raised against the Milankovitch theory. Some meteorologists pointed out that the theory dealt only with the earth's radiation balance, and ignored the role played by the atmosphere and the ocean in the transport of heat. Others found inconsistencies in some of the calculations published by Milankovitch. In theory, temperatures during an ice-age summer would have been 6.7° C colder than today. Milankovitch's calculations here seemed quite reasonable. But winter temperatures were calculated to have averaged 0.7° C *warmer* than today—a value that many scientists found difficult to believe. Milankovitch himself was untroubled by these criticisms: "I do not consider it my duty to give an elementary education to the ignorant, and I have also never tried to force others to accept my theory, with which no one could find fault."

In spite of the theoretical objections raised by meteorologists and the damaging field evidence found by Schaefer, the majority of scientists continued to favor the astronomical theory as late as 1950. But the early 1950s saw a dramatic about-face. By 1955, the astronomical theory was rejected by most geologists. The downfall of the theory was the development of a revolutionary new approach to the problem of dating Pleistocene fossils.

The new technique was the radiocarbon dating method, developed between 1946 and 1949 by Willard F. Libby at the University of Chicago. Libby discovered that a radioactive form of carbon (radiocarbon) is produced in small quantities in the atmosphere by cosmic rays. Eventually, the radiocarbon atoms in the atmosphere are absorbed into the bodies of all living plants and animals. But organisms continue to acquire radiocarbon only as long as they live. After death, the radiocarbon atoms in the organic tissues disintegrate, changing into inert atoms of nitrogen at a rate that can be measured. Libby reasoned that it should be possible to use this rate to calculate the time of death for any fossil: all that was necessary was to measure what proportion of carbon atoms in the fossil were still radioactive. Libby tested his idea extensively and found that the radiocarbon dating method worked remarkably well. The only hitch was that the dates calculated were accurate only for fossils that were less than about 40,000 years old.

When the radiocarbon dating method became available to geologists in 1951, they lost no time in launching a worldwide effort to discover the true chronology of the last ice age—or that part of it which was within the radiocarbon dating range. Radiocarbon laboratories were installed in many institutions, including Yale University, Columbia University, the U.S. Geological Survey, and the University of Groningen in the Netherlands. Pioneering geochemists such as Hans Suess, Meyer Rubin, and Hessel DeVries stood ready to analyze the anticipated avalanche of material. They did not have long to wait. Samples of wood, peat, shells, and bones were gathered from drift sheets, terrace gravels, and lake beds all over the world. "If it's organic, collect it and date it," was the rule of the day. So many dates were obtained that a special periodical, *Radiocarbon*, was established to make the results widely available.

One of the first American geologists to advocate the systematic use of the radiocarbon method in the study of Pleistocene drifts was Richard F. Flint at Yale University. After collecting a large number of datable materials from the Wisconsin drift of the eastern and central United States, Flint sent them off to Meyer Rubin for radiocarbon analysis. Flint's results showed that the drift actually recorded at least two glaciations—perhaps more. Previously, it had been supposed that a single glaciation was responsible for the Wisconsin drift, but the radiocarbon results made it clear that this hypothesis could no longer be maintained. The older tills in the drift were, for the most part, beyond the range of radiocarbon dating; but the youngest till was well within the datable range, and Flint and Rubin were able to show that the great ice sheet had reached its maximum extent 18,000 years ago. Then, about 10,000 years ago, it rapidly disappeared.

For a time it seemed that the results of the radiocarbon revolution were consistent with the Milankovitch theory. Although it was true that the 18,000-year date for the last glacial maximum was 7000 years younger than the 25,000-year date calculated by Milankovitch for the last radiation minimum, such a discrepancy could easily be explained as the time needed for a sluggish ice sheet to respond to a change in the earth's radiation budget. In fact, Milankovitch himself had predicted that just such a lag should occur, and estimated its duration as about 5000 years.

However, the discovery of a 25,000-year-old peat layer in Farmdale, Illinois, finally shattered belief in the Milankovitch theory. Such a deposit could only have been formed during an interval of relatively warm climate. Exactly how warm was uncertain, but the date for that warm interval coincided exactly with the date of a radiation minimum. When deposits of the same age and type were found at other locations in the Midwest, in eastern Canada, and in Europe, the geological evidence against the astronomical theory seemed to be overwhelming.

The program of radiocarbon dating allowed more and more geologists to fix their field observations on a firm time scale. This led to the development of a new method for constructing a climatic curve that could be directly compared with the radiation curve. Geologists accomplished this by finding dates for a large number of till and loess samples along some convenient north-south line. This till-loess boundary could then be graphically represented as a function of time. The resulting jagged line showed the position of the southern margin of the ice sheet as it advanced and retreated at that particular longitude over the course of thousands of years.

The temptation to use radiocarbon dates beyond the reliable range of the method (40,000 years) was hard to resist. By the mid-1960s, several teams of researchers had drawn diagrams showing how the southern margin of the ice sheets fluctuated back and forth during the past 70,000 or even 80,000 years. One of the most detailed of these diagrams, produced by Richard P. Goldthwait, Aleksis Dreimanis, and their colleagues was based on observations of till and loess along a line between Indiana and Quebec (Figure 31). These results revealed a pattern of climatic change that was at variance at almost every point with the astronomical theory. About 72,000 years ago, for example, the glacial margin was located in southern Quebec, far to the north of its position during the maximum advance. Yet this was the time of an important radiation minimum. Moreover, the diagram indi-

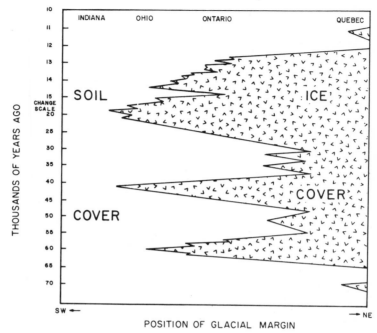

Figure 31. Fluctuations of the ice-sheet margin between Indiana and Quebec. The geographic position of the margin of the North American ice sheet is recorded as a fluctuating boundary between soil deposits and glacial drift. The supposed chronology of these fluctuations, provided by radiocarbon dates as old as 70,000 years, is at variance with the Milankovitch theory. (Adapted from R.P. Goldthwait et al., 1965.)

cated that major glacial advances occurred 60,000, 40,000, and 18,000 years ago. Only the youngest of these advances had been predicted by Milankovitch.

Wherever geologists used the radiocarbon dating method to study the older drift deposits, the result was the same. There were more glacial advances during the past 80,000 years—or at least during an interval of time they believed to the the last 80,000 years—than could be explained by the Milankovitch theory. By 1965, the astronomical theory of the ice ages had lost most of its supporters.

10

The Deep
and the Past

If the Milankovitch theory had been argued in a court of law, a
motion to declare a mistrial would certainly have been in order on
the grounds that the prosecution's case was based solely on evidence collected from the surface of the land. Since the sedimentary record of past climate on land was fragmentary, the witnesses
who testified at the Milankovitch "trial" were not only biased, they
were also poorly informed.

James Croll had been among the first to recognize the incomplete nature of the geologic record of climate, and he had anticipated the day when geologists would be able to obtain a more
complete record of the ice-age sequence by delving into the sea
bottom. "In the deep recesses of the ocean, buried under hundreds of feet of sand, mud, and gravel, lie multitudes of the plants
and animals which . . . were carried down by rivers into the sea.
And along with these, must lie skeletons, shell, and other exuviae
of the creatures which flourished in the seas of these periods."
Croll's description, however, had been purely speculative. For
during most of his life, scientists actually knew more about the
surface of the moon than they did about the depths of the sea.

But the sea would not keep her secrets for long. In 1872 the
British government equipped a 2,306-ton steam corvette, *H.M.S.
Challenger*, to make a round-the-world voyage of discovery that
would last three-and-a-half years. Under the direction of
C. Wyville Thomson, the *Challenger*'s six scientists developed
techniques for taking soundings, collecting samples of water,
capturing plants and animals, and dredging the bottom at all
depths. When the expedition returned to England in 1875, much
of the sea's mystery had been dispelled.

Observations made by the crew of the *Challenger* confirmed
many of Croll's predictions. With the exception of a few bare

ledges of basaltic rock, the sea floor was covered with a blanket of sediment. Along the margins of the continents, deposits transported to the ocean by rivers had been redistributed by currents. There, the sea bottom was covered with layers of sand and mud that contained fragments of plants and other materials derived from the land. But away from the continental margins much of the deep ocean floor was covered with fine-textured oozes. When the *Challenger*'s geologists examined samples of these oozes under the microscope, they found that many were almost entirely composed of the fossilized remains of minute animals and plants. It was not long before the living source of these tiny fossils was located. The nets that the expedition's biologists dragged through the surface waters of the ocean collected countless numbers of floating organisms (collectively known as plankton) whose mineralized remains were identical to those found on the sea floor. Clearly, the organic oozes had been formed over a long period of time by the slow rain of skeletons onto the sea floor.

Scientists on board the *Challenger* found that one type of ooze—composed of the limy remains of planktonic animals called forams—covered vast areas of the sea floor. This type of sediment was especially prevalent in temperate and tropical seas where depths did not exceed 4000 meters. Another kind of organic ooze was widespread in the colder waters of the Arctic and Antarctic seas. This type was primarily composed of opal, a glassy mineral extracted from seawater by planktonic animals called radiolaria, and by planktonic plants called diatoms. When the expedition ended, and the data that had been gathered were compiled into maps, the scientists found that the two types of organic ooze covered one-half of the sea floor—an area equal to that of all the continents combined. However, in the deepest parts of the sea, where depths exceeded about 4000 meters, the bottom was covered not with organic oozes but with a layer of brown clay that contained no fossils at all. At these great depths, it was explained, the properties of seawater were such that the opaline or limy skeletons were dissolved as fast as they rained down. All that remained here were fine particles of clay that had been wafted by currents or carried by winds.

After the *Challenger* expedition returned, an international team of investigators was organized by British scientist John Murray to analyze the staggering body of observations that had been made. By 1895 the analysis was completed and a 50-volume report published. Of particular interest to students of ancient climate was evidence that some species of forams (and other planktonic organisms) lived only in cold waters, while others lived only in warm waters. Thus, Croll's dream of extracting a complete record of climatic history from the layers of sediment on the sea floor was at last within reach. For as climate changed, the geographic extent of temperature-sensitive species would shift accordingly. The sequence of sedimentary layers at any given location would contain a permanent record of the ice-age succession.

Only one problem remained. Before scientists could reconstruct the history of climate from the succession of fossils deposited on the sea floor, they had to find a way of obtaining a cross-section of the sediment pile. Many attempts to obtain such a cross-section were made. The principle behind them was the same: a hollow steel pipe, forced into the sea bed, would extract a sediment "core" when it was removed. The earliest of these devices were called gravity corers, because the pipe was propelled into the sea floor by the force of gravity alone. A coring rig was lowered until it hung above the sea floor and then was released. The momentum gathered during its descent forced the pipe into the pile of sediment. Unfortunately, such devices were only able to collect cores about 1 meter long—not long enough to show the complete ice-age succession. To increase the depth of penetration, lead weights were attached to the coring rig, but little was gained because the frictional forces resisting penetration were too great. Other devices were tried, including an unusual one designed by Charles S. Piggot, which used dynamite to drive the pipe into the sea bottom. The method proved unsatisfactory, however, for the fossil record was badly distorted.

Despite the limitations of gravity coring, a German paleontologist named Wolfgang Schott agreed to examine a group of cores that had been raised from the floor of the equatorial Atlantic Ocean by the German *Meteor* expedition of 1925–27. Schott's results, published in 1935, set the pattern for future work on Pleistocene plankton. He began by mapping the distribution patterns of 21 different species of planktonic forams (minute surface-dwelling animals) on the present sea bed. Then, taking samples at regular intervals along his one-meter cores, he took a census of each sample. Schott found that he could distinguish three layers. In most cores, the top 30 or 40 centimeters contained an assemblage of foraminifera very different from that which was contained in the underlying layer. The assemblage in the top layer (Layer 1) was identical to that now accumulating on the sea floor. The underlying layer (Layer 2) was composed of many of the same species, but they were present in different proportions. Thus, while Layer 1 contained mostly "warm" species of forams, Layer 2 contained a higher proportion of "cold" species. In fact, one species of foram occurred only in Layers 1 and 3, and was completely absent in Layer 2. The name of that species, *Globorotalia menardii*, would be on the lips of geologists for many years to come. For Schott came to the conclusion that the sediment layer containing no *menardii* had been deposited in the last ice age when surface waters in the equatorial Atlantic were too cold to support the species. According to this view, the *menardii*-rich Layer 1 had been deposited since the glaciers retreated. Layer 3, which also contained *menardii*, was apparently a record of the interglacial interval that had preceded the last ice age.

Schott's results whetted the appetites of paleontologists for cores longer than the gravity devices were capable of producing. After all, if Schott had found a record that extended back all the way to the last interglacial age in a core only one meter long, how much more could be learned from a core that was 10 meters long?

The coring problem was finally solved in 1947, when a Swedish oceanographer, Björe Kullenberg, arranged a piston in such a way that sediments were sucked up into a coring tube while the tube was being driven into the sea floor. Because this device routinely obtained cores 10 to 15 meters long, it made possible a new era in the investigation of climatic history.

The Swedish Deep-Sea Expedition of 1947–48 was the first to use the Kullenberg corer. Led by Hans Pettersson, a crew of scientists sailed the research vessel *Albatross* around the world and obtained long cores from every ocean basin. Cores from the Pacific were sent to Gustaf Arrhenius at the Scripps Institution of Oceanography in California. By chemically analyzing these samples, Arrhenius discovered that the concentration of calcium carbonate (lime) fluctuated cyclically: layers characterized by high concentrations of limy fossils alternated with layers having lower concentrations. To explain these variations, Arrhenius reasoned that the intensity of circulation in the Pacific might have been different during an ice age than during an interglacial interval—and that it was these changes in circulation intensity that were reflected in the varying concentrations of calcium-carbonate fossils.

Arrhenius's research demonstrated that chemical as well as paleontological evidence could be used to study Pleistocene climates—at least in the Pacific. Soon, investigators at Columbia University began measuring calcium carbonate concentrations in cores taken from the Atlantic Ocean, and found that sediments from this ocean were also characterized by calcium carbonate cycles. But these cycles were opposite from those in the Pacific cores: ice-age deposits exhibited low concentrations of lime, interglacial deposits exhibited high concentrations. Evidently, the two oceans responded differently to changing climates.

The work of Arrhenius showed that sediments in the Pacific accumulated at a very slow rate—about one millimeter per century. In one sense, this was a boon to paleontologists, for it meant that even relatively short cores contained a record of the entire Pleistocene sequence. But the slow rate of deposition was also a disadvantage, for it made it almost impossible to study the details of climatic history recorded in the Pacific cores.

In the Atlantic Ocean, however, sediments accumulated at much faster rates, generally about two or three millimeters per century, so that cores taken here could be expected to contain a more complete record of climate. Geologists therefore awaited with interest the results of an investigation of 39 long Kullenberg cores that had been raised from the bed of the Atlantic by Hans Pettersson. These cores were analyzed by three scientists at the Scripps Institution of Oceanography: Fred B. Phleger, Frances L. Parker, and Jean F. Peirson. Their monograph, published in 1953, demonstrated that the long Atlantic cores recorded at least nine Pleistocene ice ages. They also found that the process of interpreting climatic history by studying deep-sea sediments was not without its problems: for several of their cores contained shallow-water forams which had obviously been displaced in some way from environments near shore. Just how these displaced faunas—and the layers of sand that were associated with them—came to be mixed in with the particles accumulating as a slow planktonic rain, was a mystery.

Before becoming a Professor of Oceanography at Scripps, Phleger had spent several years at the Woods Hole Oceanographic Institution on Cape Cod. While there, he had hired David B. Ericson to assist him in the laboratory and on shipboard. Ericson had become convinced of the need for research on marine sediments while he was Assistant Geologist at the Florida Geological Survey. Also working at Woods Hole was a geophysicist, Maurice Ewing, who was taking the first steps along a path that would eventually lead him to important discoveries about the nature of the earth's crust underneath the oceans. In 1949, Ewing was planning an expedition to the Mid-Atlantic Ridge, and he wanted an assistant experienced in the study of marine fossils. Ericson was his man.

In 1950, Ewing took a position at Columbia University and moved to New York City, taking his cores with him. Ericson later remarked that he "went along with the cores." Soon, a group of senior scientists, technicians, and students gathered at Columbia, attracted by Ewing's research on the origin of the ocean basins. The group quickly outgrew their quarters in Columbia's Schermerhorn Hall. By good fortune, Columbia had recently acquired a country estate in Palisades, New York. The estate had been given to the university by Thomas Lamont. Ewing's group moved out to the Lamont estate, and within a few years developed the Lamont Geological Observatory into a world-renowned center for oceanographic and geophysical research.

Realizing the potential importance of the core studies, Ewing insisted that Lamont vessels take piston cores every day, no matter what other research activities they were engaged in. Hundreds of cores were raised each year and stored for future study. The Lamont core collection soon became the largest in the world, and Ericson found himself in an ideal position to study the history of climate. Already familiar with the work done by Schott and by the Scripps group, he was eager to expand their findings and to work out the history in greater detail. But in some areas of the ocean, layers of displaced sediment similar to those which had stumped Phleger posed a serious problem. These layers of sand and shells—transported in some way from shallow coastal waters— distorted the climatic record produced on the deep-sea floor by the slow rain of planktonic particles.

In 1952, while Phleger's monograph was still in press, the riddle of the displaced layers was solved by two of Ericson's colleagues at Lamont. By investigating records of an earthquake that had occurred on the Grand Banks of Newfoundland in 1929, Bruce C. Heezen and Maurice Ewing were able to identify the process by which displaced layers form. The 1929 earthquake had triggered a sediment slide on the ocean floor. Particles of sediment, suspended in a turbid layer of water near the bottom, had flowed downslope. Moving with the speed of an express train, this turbidity current had broken submarine telephone cables and spread a layer of sand and mud over a wide area, disrupting the normal sedimentation process in deep water.

Now that Ericson understood what had caused the displaced layers, he was able to devise methods for identifying them, and thus remove the sedimentary static from the climatic signal. He and his assistant, Goesta Wollin, began examining every core in the Lamont library—no mean task, since they were now being collected at the rate of 200 cores per year. To speed up the process, Ericson followed a simplified version of the laboratory procedure originally developed by Schott. Instead of counting the number of individuals representing every foram species in a sample, Ericson and Wollin concentrated their attention upon the few species they considered to be particularly sensitive to changes in climate (Figure 32). Originally, they simply estimated the abundance of these indicator species. Later, as more precise results were demanded, actual counts were made. In lower latitudes, the prime candidate for this monitoring role was *Globorotalia menardii*, the species that Schott had found exclusively in his two warm layers. Ericson's work in cores taken from low latitude sites in the Atlantic confirmed Schott's idea, for here fluctuations in *menardii* abundance provided a clear record of changing climate. But in cores taken from higher, colder latitudes, *menardii* never occurred at all. In these latitudes, other species had to be used to monitor changes in past climate.

By 1956 Ericson was convinced that his simplified climatic method was valid, and he could point to supporting evidence from two different lines of research. One line was that followed by his colleagues at Lamont, Wallace S. Broecker and J. Laurence Kulp. When these geochemists dated the boundary between Ericson's two uppermost sediment layers—the upper one containing *menardii* specimens and the lower one containing no such specimens—they found that the transition occurred abruptly about 11,000 years ago. This date was very close to the radiocarbon age found for a sudden change in temperatures on land. In an article published in 1956, Ericson, Broecker, Kulp and Wollin concluded: "The core data point definitely to the period immediately before and after 11,000 years as a very critical period in glacial history. Further correlation of events both in the ocean and on land during this interval may lead to an understanding of some of the factors causing glaciation." The collaboration that

Figure 32. A fossil from the deep-sea floor. Upon death, the mineral remains of many surface-dwelling animals and plants fall to the sea floor and build up thick deposits of sediment. The shell shown here is that of *Globorotalia menardii*, a species of planktonic foram used extensively by D.B. Ericson as an indicator of Pleistocene climate. The specimen is about one millimeter wide. (Courtesy of A. Bé.)

produced this article reflected an emerging pattern of interdisciplinary study that would eventually become the hallmark of Lamont research.

The other investigation, which seemed at first to provide an independent confirmation of Ericson's climatic results, was the development of a different method for estimating the temperature of Pleistocene oceans. Developed in 1955 by Cesare Emiliani at the University of Chicago, the method was based on the isotopic composition of oxygen atoms in fossil forams. When the two methods were applied to the same cores, the results agreed quite well over the more recent part of the record. Over the older parts of the record, however, the results did not agree—a fact which would be the subject of a great deal of debate in the years to come.

By 1961, Ericson had studied more than a hundred cores, and was ready to generalize his scheme of climatic history. To facilitate discussion, he coined a set of terms that would eliminate the need for such convoluted phrases as "The third zone from the top lacking *Globorotalia menardii*." The simplified system (suggested by a harried editor working on the manuscript of the 1961 article by Ericson, Ewing, Wollin, and Heezen) used the letters of the alphabet to denote layers in the core. Thus the warm zone at the top of the cores became known as Ericson's Z Zone, representing postglacial time. The Y Zone represented the last major glacial advance; and the X Zone was the preceding interglacial, with temperatures similar to those of today (Figure 33). The new scheme was quickly adopted, making it easier to refer to an important characteristic of Ericson's climatic curve: the V Zone, which contained high concentrations of *menardii*, was unusually long. The underlying U Zone, without *menardii*, was unusually short. Ericson pointed out that his long V Zone compared well with the Great Interglacial recognized by Penck and Brückner in their study of European climate. Although Ericson himself did not support the Milankovitch theory, scientists who did were able to draw comfort from the *menardii* curve.

In the meantime, however, Ericson had become acutely aware of a conflict between the isotopic temperature method used by Emiliani and his own fossil scheme. In an attempt to resolve the conflict, Ericson and Emiliani analyzed samples from the same three Caribbean cores—each using his own method (Figure 33). Emiliani's method provided estimates in degrees Celsius, while Ericson's scanning method revealed only general temperature trends. Following the system developed by Arrhenius, Emiliani numbered his inferred temperature variations from the top down. As discussed in Chapter 11, the two methods produced broadly similar patterns for the intervals W-Z in Ericson's scheme (Stages 6-1 in Emiliani's). Only when the pattern was examined in greater detail did disturbing differences begin to be revealed. Ericson's X Zone was shorter than Emiliani's Stage 5, and many of Emiliani's cores showed a short but distinct warm interval in Stage 3 that had no counterpart in Ericson's Y Zone. Moreover, Ericson's unusually long V Zone showed up in Emiliani's scheme as several separate fluctuations. And Ericson's U Zone (cold) de-

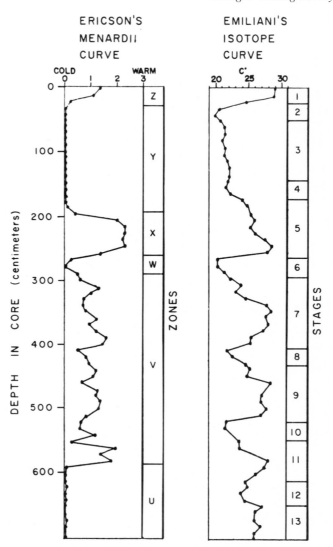

Figure 33. Succession of Caribbean ice ages according to Ericson and Emiliani. Fluctuations in the abundance of *Globorotalia menardii* down a deep-sea core from the Caribbean (A179–4) were interpreted by D.B. Ericson as a climatic record. Cold intervals were referred to as Zones U, W, and Y; warm intervals as Zones V, X, and Z. Measurements of the oxygen-isotope ratio made by C. Emiliani from samples of the same core were also interpreted as a climatic record. Warm intervals were referred to as Stages 13, 11, 9, 7, 5, 3, and 1; cold intervals as Stages 12, 10, 8, 6, 4, and 2. (Data from C. Emiliani, 1955 and D.B. Ericson et al., 1961.)

parted dramatically from Emiliani's warm Stages 11 and 13. No satisfactory explanation for the divergence between the two schemes was to be found for almost a decade.

By 1963, Ericson, Ewing, and Wollin had made great strides toward their goal of charting the climatic history of the Pleistocene Epoch. After analyzing "more than 3000 cores raised from all the oceans and adjacent seas during 43 oceanographic expeditions since 1947 . . . we have found eight containing a boundary clearly defined by changes in the remains of planktonic organisms." This boundary, they concluded, marked the onset of the first ice age of the Pleistocene. The boundary to which they referred marked the extinction of a group of star-shaped fossil plants called discoasters, an event which they estimated had occurred 1.5 million years ago.

But geologists in general remained unconvinced of the correctness of the conclusions reached by Ericson and his colleagues. They questioned the chronology, because that was based on assumptions that were difficult to prove. And they questioned the climatic interpretation Ericson had given to the sequence of *menardii* zones. In 1964, Ericson and Wollin explained their methods in a book, *The Deep and the Past*. Yet skepticism remained. After all, Emiliani had obtained quite different results—and derived them from sediments of the same ocean, and from some of the very same cores.

11

Pleistocene Temperatures

In 1949, when Ericson and Ewing were raising their first cores
from the Mid-Atlantic Ridge, Cesare Emiliani was on his way to
the University of Chicago to begin postgraduate studies in
paleontology. He brought with him several rock samples from the
northern Appenines in Italy, where he had been working as an oil
geologist since his graduation from the University of Bologna in
1945. Emiliani planned to spend about a year at the University of
Chicago, broadening his intellectual horizons and investigating
some intriguing aspects of his Appenine samples.

At the University of Chicago, Emiliani met several young scientists who were working with Nobel laureate Harold C. Urey to
find geochemical answers to many of the fundamental questions
of earth history. One of these young scientists, Samuel Epstein,
later remembered: "When I arrived in Chicago, the place was in a
ferment. New ideas popped up every day. Not only did we have
Harold Urey for inspiration, but there were other intellectual
stars as well, including Willard Libby and Enrico Fermi." (Libby
later received the Nobel prize in chemistry; Fermi was already a
Nobel laureate in physics.)

Epstein was helping to develop an idea that Urey had originated in 1947. Urey theorized that it should be possible to use
oxygen atoms to find out what the temperature of the ocean had
been in the past. The technique was based on the fact that sea
water contains two distinct types (isotopes) of oxygen atoms. One
of these isotopes (oxygen-18) is heavier than the other (oxygen-
16). Both types are present in the calcium carbonate skeletons of
marine organisms. Urey had demonstrated theoretically that the
amount of the heavier isotope that an animal extracts from the
surrounding sea water depends on the temperature of the water.
In cold water, the skeletons have higher concentrations of the
heavier isotope. Therefore, he reasoned, it should be possible to
calculate the water temperature during the lifetime of the or-

ganism by measuring the ratio of oxygen-18 to oxygen-16 in a fossil skeleton.

Epstein and Emiliani were among the researchers in Chicago who were convinced that Urey's method would provide valuable insights into the history of the earth. Before they could proceed further, however, they had to overcome two obstacles. The first was theoretical: the ratio of isotopes in the skeletons was influenced not only by the temperature of the water but also by the isotopic composition of that water. If that composition varied, it would be impossible for researchers to interpret temperatures accurately. But Urey and his colleagues were confident—overconfident, as events were to prove—that a solution to this problem would be found as their investigation progressed. The second obstacle was a technical one: instruments and laboratory procedures had to be developed that would provide researchers with extremely accurate isotopic data. This was the problem to which Epstein and his coworkers, Ralph Buchsbaum and Heinz Lowenstam, were addressing themselves when Emiliani arrived on the scene. After several years of work, Epstein's group succeeded in developing a laboratory procedure for measuring isotopic ratios with great accuracy. The way was open to investigate the potential of Urey's temperature method.

Urey soon realized that Emiliani's knowledge of fossils would prove useful in applying the new geochemical techniques. In 1950, he asked Emiliani if he would be interested in investigating the isotopic composition of foraminifera. Without hesitation, Emiliani accepted—here was a once-in-a-lifetime chance to open a new window on the geologic past. The other members of the Chicago team were already using the "isotopic thermometer" on fossils from remote geological periods, but Emiliani planned to apply the method to skeletons of foraminifera in Pleistocene deposits. His first measurements were made on bottom-dwelling forams collected from deposits in California in 1951. But when Hans Pettersson offered him eight piston cores raised by the *Albatross* expedition, Emiliani decided that a study of planktonic foraminifera had even greater potential. Soon, Ewing put several Lamont cores at his disposal; and Ericson, eager to verify his own climatic method, mailed to Chicago samples of four cores he had already analyzed.

By August 1955, Emiliani had analyzed eight deep-sea cores. His conclusions were published in an article called "Pleistocene

Temperatures," which appeared in the *Journal of Geology* (1955). The article proved to be a landmark in the study of the ice ages. According to Emiliani, isotopic variations in cores taken from the Caribbean and from the equatorial Atlantic indicated that over the last 300,000 years there had been seven complete glacial-interglacial cycles. The data also seemed to indicate that during a typical ice age, surface waters in the Caribbean dropped approximately 6° C (Figure 33). Finally, Emiliani noted that the variations in isotopically estimated temperatures showed a reasonably good time correspondence with the Milankovitch radiation curves, and concluded that his observations tended to support the astronomical theory of the ice ages.

With the publication of his article, Emiliani found himself embroiled in three separate arguments. The first was with Ericson. Did the variations in the isotopic curve really reflect variations in temperature? Or did the variations in Ericson's *menardii* curve reflect the history of climate more accurately? Second, Broecker and Ericson questioned the accuracy of Emiliani's time scale. The third and last argument came from the many geologists who had rejected Milankovitch's theory. They held that the observed correlation between Emiliani's isotopic curve and Milankovitch's radiation curve was merely a coincidence.

In many respects, the first of these arguments was the most vital. If Ericson was right, and the isotopic variations were caused by something other than temperature changes, then the other two arguments lost much of their point. By 1964, the importance of finding a solution to the Ericson-Emiliani controversy was widely recognized. Wallace Broecker of Columbia University joined with Richard F. Flint and Karl Turekian of Yale in persuading the National Science Foundation to sponsor a conference on the problem. The goal of the conference would be to resolve the issue by having both Ericson and Emiliani present their ideas and the supporting data before a small group of specialists. In January 1965, a two-day conference was held at the Americana Hotel in New York City, and at Columbia's Lamont Geological Observatory.

Among those in attendance was John Imbrie, then Professor of Geology at Columbia University. For more than a decade, Imbrie had been studying bottom-dwelling fossils and using them to interpret the climates of remote geological periods. He had adapted a statistical method, called factor analysis, which had proved to be helpful in studying how marine animals respond when they are subjected to a variety of environmental stimuli simultaneously.

The long-awaited Ericson-Emiliani debate of 1965 was inconclusive. On the one hand, Ericson demonstrated that his zonation scheme applied to hundreds of Atlantic cores and presented evidence that his main indicator species (*menardii*) was indeed sensitive to changes in ocean temperature. He also criticized Emiliani's assumption that changes in the isotopic composition of the ocean during the Pleistocene had been so small as to have little effect on his paleotemperature estimates. In fact, Ericson argued, many isotope chemists had drawn just the opposite conclusion— that ice sheets contained such high concentrations of the light oxygen isotope that significant changes in the isotopic composition of the ocean must occur during the ice-age cycle. It was, therefore, quite possible that the isotope variations measured by Emiliani were not connected at all with ocean temperatures, and only reflected changes in the volume of the ice caps.

For his part, Emiliani criticized Ericson for relying mainly on only one species, and he presented data (gathered by Louis Lidz) that supported the oxygen-isotope results. Lidz had studied many different foram species in two of Emiliani's cores and had found that fluctuations in their abundance correlated well with fluctuations in the isotope curves. Emiliani also argued that the Pleistocene ice caps were probably not as rich in oxygen-16 as geochemists thought, and reiterated his belief in the accuracy of his temperature estimates.

Although he was more an onlooker than an active participant in the debate, Imbrie pointed out that Ericson and Emiliani were both ignoring the possibility that factors other than temperature might have operated to cause variations in foram concentration. For example, changes in the salt content of the water, or in the amount of available food, would certainly have an influence on foram populations. He then went on to suggest that the application of statistical techniques to the entire assemblage of foram species should make it possible to separate the temperature effect from other environmental influences. Before the conference ended, Imbrie had decided to make this attempt himself.

Ericson was ready with help and advice, and he suggested that Imbrie should investigate one particular core in detail. This core (already analyzed for *menardii* by Ericson and Wollin) was known as V12–122 because it had been taken by the Lamont research vessel *Vema* on the 12th cruise and 122nd station. Imbrie found an able associate in Nilva Kipp, an undergraduate at Columbia's School of General Studies, who had written an impressive term paper on the Ericson-Emiliani controversy. Working together over the next three years, first at Columbia and then at Brown University, Imbrie and Kipp developed a multiple-factor method for climatic analysis that took into account abundance variations in all 25 species of planktonic forams. In many respects, their approach was a computerized extension of the technique used by Wolfgang Schott in 1935. Schott's first step had been to show the present distribution of each foram species on a series of maps. Imbrie and Kipp followed this procedure, and then went on to write equations to express the relationship between the species abundances on the sea floor and various properties of the surface waters. These properties included summer and winter temperatures and salinity (salt content). Working down through the core, the equations—which had been developed for the present sea floor—were used to estimate summer and winter temperatures and salinity for past epochs.

By the summer of 1969, Imbrie and Kipp were sure that their multiple-factor technique produced dependable results. In the meantime, Broecker and Jan van Donk had conducted an isotopic analysis of the same core (V12–122), making it possible to compare the results of the methods developed by Ericson, by Emiliani, and by Imbrie and Kipp. The comparison convinced Imbrie and Kipp that Ericson had been wrong—and Emiliani, half right. For where Ericson's zones showed cold temperatures, both the isotope and multiple-factor methods showed warm intervals. Apparently, some environmental factor other than surface water temperature (but often correlated with it) caused *Globorotalia menardii* to appear and disappear cyclically in deep waters of the Atlantic Ocean.

On one fundamental point, however, the results of the multiple-factor technique did not mesh with Emiliani's. Imbrie and Kipp's research showed that the temperature of Caribbean surface waters had dropped only 2° C as the world entered each ice age—not the 6° C postulated by Emiliani. The multiple-factor method showed that changes in the salinity of surface waters in the Caribbean had affected the foram population in that area along with changes in temperature. By attributing all of Lidz's faunal variations to temperature, and ignoring other influences, Emiliani had overestimated the magnitude of the temperature change. If Imbrie and Kipp were right in their estimate of a 2° C lowering of Caribbean temperatures during ice ages, an important conclusion could be drawn: much of the isotopic variation must be due to changes in the volume of the ice sheets—not to changes in temperature.

Eager to announce their results, Imbrie was elated when Emiliani invited him to speak at an international scientific meeting to be held in Paris in September 1969. But he arrived late, and his lecture had to be rescheduled—for four o'clock on Friday. In Paris, on a warm September afternoon, there are distractions attractive enough to lure even the most dedicated of scientists away from a lecture hall. When Imbrie finally spoke, it was to an audience of two. Half of the audience understood no English. The other half was Nicholas Shackleton—a young British geophysicist who, unknown to Imbrie, had already published data suggesting that much of the observed isotopic variation reflected changes in the volume of global ice.

Meeting after the lecture, Imbrie and Shackleton were delighted to find that their independent approaches to the study of climate history had led to the same tentative conclusion. Although they realized that many more cores would have to be analyzed before they could be certain, the available data seemed to indicate that fluctuations in Emiliani's isotope curve were recording primarily variations in the total volume of the ice sheets.

To some scientists, perhaps even to Emiliani himself, such a result might be a disappointment. Hopes had been high that Urey's geochemical method would provide a method for estimating the temperature of Pleistocene oceans. But to the two men in Paris, it seemed that the importance of the isotope curve would be greatly enhanced if its role as a device for measuring global ice volume could be firmly established. After all, what could be more useful in the analysis of Pleistocene history than to know how the size of the ice sheets varied with time? Given the isotope technique to record the volume of ice, and the multiple-factor technique to record the temperature of the ocean, a way might yet be found to test some of the competing theories of the Pleistocene ice ages.

12

Milankovitch Revival

By 1969, the majority of scientists were sufficiently impressed with the radiocarbon evidence against the Milankovitch theory to eliminate the idea as a serious contender in the ice-age sweepstakes. Only a minority continued to search for methods of testing its validity. Among that minority was Rhodes W. Fairbridge, a geologist who had made a detailed study of ancient sea levels. Fairbridge was particularly impressed with evidence he had found along the south coast of Australia, where nineteen parallel ridges of sand marked the position of former shorelines, and recorded times when sea level was higher than today. Although there was no way of knowing how old these abandoned shorelines were, the fact that they were evenly spaced strongly suggested that the ups and downs of sea level occurred in a regular rhythm. Knowing that oscillations of sea level were caused by the melting and growth of ice sheets, Fairbridge inferred that the ice ages themselves must have recurred at regular intervals. To Fairbridge, these facts and inferences lent strong support to the astronomical theory: "The mechanism proposed by Milankovitch struck me as reasonable," he said later, "and the astronomical cycles that were the basis of his theory were about the right length to explain the sand ridges along the Australian coast."

Although this argument was plausible, Fairbridge's critics were quick to point out that it was purely qualitative, and that there was no way of knowing what the length of his sea-level cycle really was. Did it correspond to the 41,000-year cycle of axial tilt, the 22,000-year cycle of precession, or some entirely different periodicity? If a way could be found to date the episodes of high sea level, then it would be possible to use that information as a test of the Milankovitch theory. Unfortunately, all of the shoreline features studied by Fairbridge were older than 40,000 years, and thus well beyond the effective range of the radiocarbon dating technique.

But a second revolution in dating techniques was already well underway. Geochemists in several laboratories were working to develop dating methods that depended not on radiocarbon, but on radioactive isotopes of uranium, thorium, and potassium. Eventually, ten different techniques were developed whose accuracy depended on the age and type of material being dated. The potassium-argon method, for example, gave accurate results on volcanic rocks of almost any age; but another method, based on a radioactive substance called protactinium, yielded only approximate results and could be used only on deep-sea muds younger than about 150,000 years.

In 1956, John W. Barnes and his colleagues at the Los Alamos Scientific Laboratory had developed a dating method using thorium that could be used to obtain accurate dates on ancient coral reefs, provided that the reefs were no older than about 150,000 years. It was this method, therefore, that had the greatest potential for determining the chronology of ancient sea levels, and it was this method that would soon provide the first critical test of the Milankovitch theory.

One of the leaders of the new dating revolution was Wallace S. Broecker. Since coming to Columbia University as a graduate student in 1952, Broecker had devoted himself to improving the geological calendar. Trained in geochemistry, his first work had been the application of the radiocarbon method to the dating of late Pleistocene events. His research had helped to persuade geologists that, around 11,000 years ago, there had been a short interval of rapid climatic change. This change had caused the level of lakes in the arid regions of the southwest to rise; the ice sheets to recede; and sea level to rise. By the early 1960s, Broecker and his students were improving the thorium method and using it to date levels of the sea old enough to be well beyond the range of the radiocarbon method.

In August 1965, Broecker reviewed the existing knowledge of sea-level history in a lecture given at an international scientific congress held in Boulder, Colorado. It should now be possible, he said, to establish some firmly fixed points in what had been a rather flexible geological time scale. There were already important results to report since he and his student, David Thurber, had dated ancient fossil reefs in Eniwetok atoll, the Florida Keys, and the Bahama Islands. The results indicated that about 120,000 years ago the sea stood about six meters higher than it does now. One sample seemed to indicate that another high stand of the sea occurred about 80,000 years ago. Exactly how much higher was not known. Broecker drew a diagram showing that the three known dates for high sea levels—today, 80,000 years ago, and 120,000 years ago—corresponded reasonably well with three of the four maxima of Milankovitch's radiation curve for 65° N. Although admitting that this analysis was only preliminary, Broecker expressed his conviction that definitive information would become available within the next few years.

In the spring of the same year that Broecker gave his lecture, Professor Robley K. Matthews of Brown University boarded a plane for the tiny Caribbean island of Grenada. His fellow passengers were looking forward to a holiday in the sun, but for Matthews this trip was all business. A specialist in limestones, he was particularly interested in the processes by which these rocks may become porous enough to serve as reservoirs for oil. Having heard that limestone outcrops on Grenada might be profitably studied, Matthews had decided to investigate.

Matthews spent the first night of his journey on the island of Barbados, and was happy to find that there, at least, limestone was widely exposed. But his first view of Grenada showed him that he had been misinformed. The island was a pile of volcanic rock, with almost no limestone at all. Upon landing, Matthews promptly reserved a seat on the next plane out.

Back on Barbados, Matthews discovered that much of the island was terraced, so that from the air it resembled a huge flight of stairs. Although limestone exposures were poor on the flat "treads" of these terraces, they were excellent on the steep "risers." Matthews returned to his teaching duties at Brown satisfied that he had found an appropriate location for his field studies.

He soon discovered that there was considerable controversy over how the Barbados terraces had been formed. According to one theory, the island periodically rose out of the sea. Each time the island rose, one reef was killed, and another developed along the new shoreline at a lower position on the island. Each terrace, therefore, represented a separate episode of reef growth at a particular sea level. Another theory held that the terraces had been carved out of a single large fossil reef by the action of waves. As the island rose, the argument went, erosion carved a new terrace at each succeeding elevation.

Matthews resolved to find out whether the Barbados terraces were accretionary or erosional. Later that summer he returned to the island with a graduate student, Kenneth Mesolella. Along each terrace they found outcrops exposing a cross-section of an ancient coral reef. In some places tree-like colonies of the coral *Acropora palmata* were still standing in the positions they had assumed when alive. Only a few yards offshore, living individuals of the same species were building shallow-water reefs similar to the fossil ones. By the end of the summer, Mesolella had studied every limestone outcrop on the island, and he and Matthews became convinced that accretion was responsible for the sequence of Barbados terraces—and that each terrace represented reef growth at one former level of the sea. To simplify discussion, they numbered the terraces in order of elevation, starting with Terrace I.

Another Brown University professor, Thomas A. Mutch, suggested that the reef sequence might be useful in the study of the history of sea level. Although skeptical that this objective could be achieved on an island with a history of vertical movement, Matthews nevertheless persuaded Broecker to date some of the terrace samples.

Which reefs should be dated first? Matthews decided to send samples of the first and third reefs, from Terraces I and III. Broecker, John Goddard, and graduate students David Thurber and Teh-Lung Ku set to work. By summer the first laboratory measurements had been completed. The two reefs were 80,000 years and 125,000 years old. Broecker was pleased with this result, because it agreed with reef dates already determined from the Bahamas and Florida—and because these dates matched reasonably well with the only two dates given by Milankovitch for radiation maxima in that part of the time scale.

But Broecker was shaken out of his complacency when the Brown group informed him that there was another terrace between the two he had dated. Samples were sent from this "terrace in the middle" and these gave a date of 105,000 years.

Since the radiation curve for 65° N had no maximum at 105,000 years, Broecker began examining other Milankovitch radiation curves. He soon made the important discovery that the curves for lower latitudes (particularly 45° N) contained distinct peaks near all three Barbados terrace dates: 82,000, 105,000, and

125,000 years ago. Previously, supporters of the Milankovitch theory had focused their attention on the curve for 65° N—a curve which is so strongly influenced by variations in the axial tilt of the earth that its peaks are spaced about 40,000 years apart. But at lower latitudes, the 22,000-year precession cycle has a strong enough effect to modulate significantly the effect of variations in tilt. Thus, the observed sequence of terrace dates from Barbados seemed to be telling the investigators that the precession cycle was more important than Milankovitch had believed.

These discoveries, published in 1968 and confirmed within a few years by independent investigations on New Guinea and the Hawaiian Islands (Figure 34), led to a general revival of interest in the Milankovitch theory. For Broecker, Matthews, and Mesolella had shown that the astronomical theory—if modified to place more emphasis on the precession effect—could account for the episodes of high sea level that occurred 82,000, 105,000, and 125,000 years ago (Figure 35).

Figure 34. Reef terraces on New Guinea. A view along the north coast of the Huon Peninsula, showing uplifted terraces formed by Pleistocene coral reefs. Terraces similar to these were first dated on the Caribbean island of Barbados. (Courtesy of A. Bloom.)

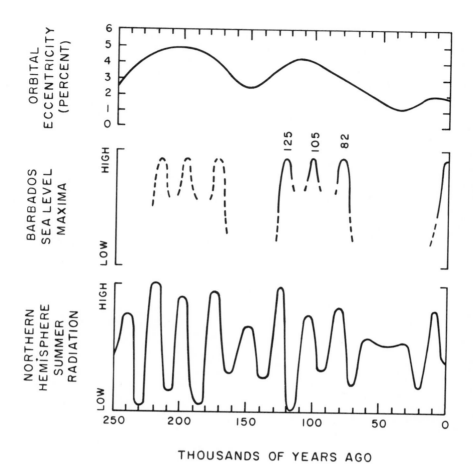

Figure 35. Astronomical theory of Barbados sea levels. Accurately dated episodes of high sea level are shown as solid lines in the middle diagram. Episodes of uncertain age are shown as dashed lines. Known times of high sea level correspond to intervals of intense summer radiation and high orbital eccentricity. (Adapted from Mesolella et al., 1969.)

But this revival of interest did not lead automatically to firm belief. As Broecker and Matthews were quick to point out, the match they had discovered between the three terrace dates and the three radiation dates could be due to chance. Perhaps a causal connection did not exist. To clinch the argument they needed a string of coincidences so long that chance alone could not be responsible. To make such a test of the astronomical theory, geologists needed a geological calendar for Pleistocene events much longer than that provided by thorium dates from ancient coral reefs.

13

Signal
from the Earth

The key that would eventually unlock the chronology of the Pleistocene had been found in 1906 in a French brickyard by Bernard Brunhes, a geophysicist investigating the earth's magnetic field. Brunhes discovered that as a newly baked brick cools, iron-rich mineral particles align themselves parallel to the direction of the earth's magnetic field, so that the brick becomes slightly magnetized. What made Brunhes's observation important to geology was his further discovery that cooling lava flows behave like bricks, acquiring a direction of magnetization parallel to the earth's field. Brunhes concluded that ancient lava flows must contain information about the history of the earth's magnetism.

Intrigued by this idea, Brunhes measured the direction of magnetization in several ancient lava flows and was astonished to discover that some of the flows were magnetized in a direction directly opposite from that of the present magnetic field. At certain times in the past, he concluded, the earth's magnetic field must have been reversed. If this were true, an observer—transported back in time to an epoch of reversed polarity—would see the north-seeking end of his compass needle swing around to point due south. This idea seemed so unlikely that few of Brunhes's contemporaries accepted it.

More than 20 years later, however, a Japanese geophysicist found evidence that Brunhes had been right. After studying a succession of lava flows in Japan and Korea, Motonori Matuyama concluded that the earth's magnetic field had reversed itself at least once during the Pleistocene Epoch. Moreover, further investigation convinced him that field reversals had occurred many times during geological epochs much older than the Pleistocene. If proven correct, Matuyama's concept of multiple field reversals would have an important impact on historical geology. For these events, recorded simultaneously in lava flows on every continent, would provide what geologists had long needed—a method of precise correlation between widely separated deposits.

But if one reversal had seemed unlikely, multiple reversals seemed bizarre, and Matuyama's work was treated with much skepticism. Before long, geologists found a less dramatic way to explain the facts. Studies showed that certain minerals, cooling under specific conditions in a laboratory, acquired a reversed polarity. If this peculiar mechanism of self-reversal operated in the laboratory, it might also have operated in ancient lava flows. Thus, in spite of the fact that minerals capable of self-reversal were rare in lavas, almost all of the scientists who considered the question of magnetic reversals found self-reversals easier to accept than the revolutionary idea that the earth's field had periodically reversed itself.

In the late 1950s and early 1960s, geophysicists working in Russia (A. N. Khramov), Iceland (Martin G. Rutten), and Hawaii (Ian McDougall and Donald H. Tarling) found evidence that Brunhes and Matuyama had been right after all—that nature had provided them with a convenient, worldwide correlation technique. Final confirmation of the field-reversal hypothesis was provided in 1963 by Allan Cox and Richard R. Doell of the U.S. Geological Survey, and by G. Brent Dalrymple of the University of California at Berkeley. To honor the memory of their pioneering colleagues, these three agreed to name the late Pleistocene epoch of "normal" polarity the "Brunhes Epoch," and the earlier epoch of reversed polarity the "Matuyama Epoch."

Cox and his colleagues proved that the field-reversal theory was correct by showing that each reversal had been a globally synchronous event. They argued that it would be unreasonable to suppose that lava flows all over the world had undergone self-reversal simultaneously. To demonstrate synchroneity, they dated lava flows occurring just above and just below a large number of reversals. This dating effort, carried out by a group of investigators at the University of California led by Garniss H. Curtis and Jack F. Evernden, was based on the potassium-argon method—a technique that worked particularly well with lava flows. The results not only established the synchroneity of magnetic reversals, but also focused attention on the reversal dates themselves. These dates proved to be the long-awaited fixed points around which a firm Pleistocene chronology could be constructed.

Before long, geologists found it necessary to adopt a graphic scheme for recording the history of the earth's magnetic field (Figure 36). Intervals of polarity like that observed today and during the Brunhes Epoch were called "normal" and recorded as a black bar. Intervals of reversed polarity, such as the Matuyama Epoch, were recorded as a white bar. Soon, two short "normal events" were discovered within the reversed epoch of the Matuyama. The older of these, the Olduvai Normal Event, was named for the valley in Africa where evidence for it was first discovered. The younger was named the Jaramillo Normal Event, for a small river in New Mexico. On a graph, the earth's alternating magnetic field appeared as a series of black and white bars—similar to a message in Morse code. Two points on the signal were of crucial importance to geologists attempting to unravel the mystery of Pleistocene ice ages. One of these points was the Brunhes-Matuyama boundary, dated at 700,000 years ago; the other was the Olduvai Normal Event, dated at 1.8 million years ago.

But would this signal be recorded in deep-sea cores? As early as 1956, Maurice Ewing and Manik Talwani attempted to find out. Talwani took two of the Lamont cores to the Carnegie Institution in Washington, D.C. where John Graham measured their magnetic properties. "We did find several reversals," Talwani said later, "but the results were ambiguous." One problem was that the cores were still very soft and difficult to handle. After making several other attempts, Talwani abandoned the effort.

Ten years later, two geologists working at the Scripps Institution, Christopher G. A. Harrison and Brian M. Funnell, found evidence of the Brunhes-Matuyama boundary in two Pacific cores. Although the Lamont investigators were not convinced by this evidence, they mounted a new effort to read the paleomagnetic signal. They had every reason to hope that Harrison and Funnell were right. For the Lamont collection contained more than 3000 long cores taken from all of the world's oceans—and locked inside each one was a wealth of climatic information waiting to be placed in the chronologic framework provided by the calendar of magnetic reversals.

Fortunately, the Lamont staff now included a specialist in rock magnetism, Neil D. Opdyke. Since Opdyke had previously

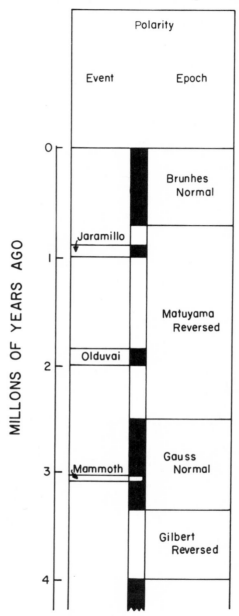

Figure 36. Magnetic history of the earth. *Epochs* and *events* of normal polarity are shown as black bars. *Epochs* and *events* of reversed polarity are shown as white bars.

worked primarily with rocks, he invited John Foster, a graduate student, to help him develop an instrument for analyzing soft sediments. Later, another student, Billy Glass, became the third member of the team, adding his specialized knowledge of ocean sediments to that of Opdyke and Foster. The trio decided to investigate cores from high latitudes, where the inclination of the magnetic field would be steeper and reversals would be easier to detect. Realizing that erosion on the sea floor might have removed parts of the sedimentary record, they also decided to analyze only those cores that had already been examined by a paleontologist.

The group asked Lamont's James D. Hays, whose specialty was Antarctic radiolaria (minute surface-dwelling animals), to select cores for magnetic analysis that would be long enough to penetrate the Brunhes-Matuyama boundary. Since his undergraduate days at Ohio State University, Hays had been fascinated by the Antarctic. In fact, he had first been attracted to Lamont by the library of cores that would enable him to study the history of the Antarctic Ocean. His first step in that direction had been to develop a detailed sequence of radiolarian zones. With such a background, Hays found it an easy task to choose cores for paleomagnetic analysis. His own radiolarian zones would provide an additional check on the paleomagnetic correlations.

When the samples had been analyzed, the four investigators were delighted to find that the magnetic signal was clearly impressed on all of the cores. Moreover, the correlations provided by the reversal boundaries confirmed those previously derived from Hays's radiolarian zones. With this decisive confirmation of Harrison and Funnell's findings, the "paleomagnetic revolution" began. For the first time, it was possible to date climatic events in any core that recorded the paleomagnetic signal. Now the value of the Lamont library of cores became apparent. Without waiting for new cores to be taken, Opdyke, Hays, Ericson, and their colleagues would between 1966 and 1969 transform what had been a generalized geological narrative into a well-dated history of climate.

The first problem to be attacked was the length of the Pleistocene Epoch. Following the lead of Penck and Brückner, Milankovitch had assumed that the Pleistocene was 650,000 years long, and confined his early calculations to that interval. Only after his death had Ericson and his colleagues at Lamont estimated that the Pleistocene was approximately 1.5 million years long. Clearly, before a satisfactory test of the Milankovitch theory could be made, geologists would have to date the beginning of the Pleistocene Epoch.

The problem was actually two-fold: What event defined the beginning of the Pleistocene, and when did that event occur? During the previous century, the first question had stimulated several answers, beginning with Charles Lyell's 1839 proposal that any deposit be labeled Pleistocene if 90 to 95 percent of the fossil species it contained were still alive today. No mention here of an ice age or cold climates. Later, Edward Forbes had substituted a climatic definition: a Pleistocene deposit was one which contained evidence of a cold climate. But how cold?

In 1948, an international committee of scientists had settled on an unambiguous—if somewhat arbitrary—definition of the Pleistocene. The beginning of the Epoch, they said, was marked by the first appearance of cold-water species in well-exposed sedimentary successions in southern Italy. But the practical problems of using this definition were great. How, for example, could a scientist studying a core from the Pacific determine which level in that core correlated with the defining event in Italy?

The problem of correlating and dating the base of the Pleistocene was solved by application of the paleomagnetic method. William A. Berggren of Woods Hole Oceanographic Institution and James D. Hays showed that the first appearance of cold-water species in southern Italy coincided with the Olduvai Normal Event. After a century of struggle geologists could at last state that the Pleistocene Epoch began 1.8 million years ago. Now they could go on to use magnetic reversals within the Pleistocene—particularly the reversal boundary that marks the base of the Brunhes Epoch, 700,000 years ago—to establish a calendar covering that portion of Pleistocene history for which Milankovitch had constructed his theory.

14

Pulsebeat of Climate

At the same time that Broecker and Matthews were studying the history of sea level, and while Hays and Opdyke were developing the paleomagnetic time scale, a geologist named George Kukla was hard at work in Czechoslovakia, digging a pit in a quarry near the city of Brno. The quarry was located on Red Hill, where deposits of wind-blown silt (loess) were extensively used in the manufacture of bricks. By examining the walls of this excavation, Kukla and his colleague Vojen Ložek had explained why the well as the intervening layers of soil.

Kukla's interest in loess was an outgrowth of his fascination with Czechoslovakian caves. In many of these caves, thin layers of loess blown in during Pleistocene ice ages have been found to contain the bones of Neanderthal and other Stone Age people. By tracing these layers of loess outside of the caves, and correlating them with the thicker sheets that cover the sides of nearby hills, archaeologists could place the human artifacts in an historical sequence.

The Pleistocene ice sheets that flowed outward from centers in Scandinavia and the Alps never reached the Red Hill region, yet the climate there changed drastically. As early as 1961, George Kukla and his colleague Vojen Ložek had explained why the nonglaciated areas of Czechoslovakia and Austria were ideally located to record the fluctuations of Pleistocene climate. When the ice sheets were large, Central Europe was a polar desert—dry, treeless, and swept by bitter winds that deposited layers of loess. But when the glaciers were small, Czechoslovakia had a climate even warmer and wetter than today's: broad-leafed trees grew in forests, fertile soils were formed, and Stone Age hunters lived under temperate conditions. Thus, as the Scandinavian and Alpine ice sheets alternately expanded and contracted, the boundary between prairie and forest marched back and forth across the nonglaciated corridor of Central Europe.

Long before they were aware of the magnetic time scale, Czechoslovakian geologists had demonstrated that at least ten repetitions of the soil-loess cycle were recorded in the region of Brno alone. But it had not been possible to determine how long each cycle was. In 1968, Kukla and his colleagues at the Czechoslovakian Academy of Science returned to their brickyards, examined each layer of soil and loess, and found five magnetic reversals. With the time scale now fixed, the average length of each cycle could easily be calculated: the main pulse of late Pleistocene climate was a steady beat of one cycle per 100,000 years.

Investigations carried out over the preceding decade had established that the sedimentary cycle was not really a simple repetition of soil (layer 1) and loess (layer 2) in a symmetrical pattern (1-2-1-2). Instead, it was a four-fold cycle made up of three kinds of soil (1, 2, and 3) and loess (4), forming a sawtoothed, asymmetrical sequence (1-2-3-4-1-2-3-4). The first soil in the sequence formed in a warm, moist climate. The second layer was a black soil, identical to that forming now in the moister parts of the Asiatic steppe, and containing fossils indicative of a climate somewhat cooler and dryer than that of the preceding forest phase. Above the black soil was a layer of brown soil, typical of the more temperate parts of Arctic regions today. This soil, the third layer in the sequence, contained fossils indicative of a climate colder and dryer than the steppe, but not as cold and dry as that which accompanied the deposition of the overlying loess sheet that formed the fourth and final phase of the cycle.

These observations led Kukla to an important conclusion: the cooling phase of the climatic cycle lasted much longer than the warming phase. Moreover, transitions from dusty, polar desert phases to deciduous forest phases were so abrupt that they appeared in the quarry walls as distinct lines. These lines, named "Marklines" by Kukla, were useful in distinguishing one sedimentary cycle from another, and in correlating the cycles between widely separated regions (Figure 37).

Having taken the pulse of Central European climate, and finding it to be regular, Kukla turned his attention to the sequence of Alpine terraces that had led Penck and Brückner to the conclusion that Pleistocene climate had an irregular rhythm, marked by the gravels they had labeled Günz, Mindel, Riss, and Würm. That the Alpine terraces were real, Kukla had no doubt. What was in question was the climatic interpretation that Penck had given them.

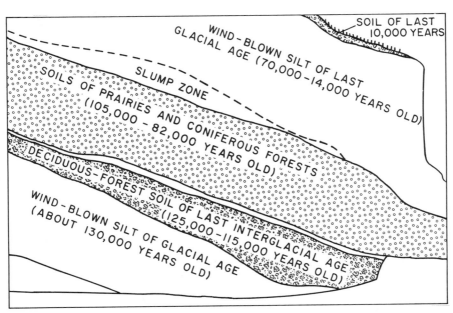

Figure 37. Climatic history recorded in a Czechoslovakian brickyard. Events of the past 130,000 years are recorded as a sequence of soils and wind-blown silts (loesses) in a quarry at Nové Město. (Courtesy of G.J. Kukla.)

Penck had assumed that gravel layers formed only during glacial times. However, Kukla soon found what he regarded as "spectacular illustrations of interglacial gravel accumulation" that confirmed the suspicions voiced by the German geologist Ingo Schaefer many years before. In the lower terrace at Ulm, for example, gravels classified as belonging to the Würm ice age contained logs that were shown by radiocarbon dating to be of postglacial age. And near Vienna, Würm-age gravels were found to contain Roman bricks. Kukla wrote:

> Yet such was the authority of Penck that the interglacial layers within the terrace bodies were interpreted as local anomalies rather than proof of the partly interglacial age of all of the Alpine terraces . . . . For instance, the gravels of the lower terrace close to Ostrava were for many years mapped as being of Würm age, which in fact they are. But after the famous Czech Quaternary stratigrapher, Tyráček, unearthed a rusty bicycle steering rod from the intact gravel, this level was recharted as Holocene alluvium . . . . Nevertheless, the logical conclusion escaped: the gravels of the lower terrace are by definition Würm gravels, whether or not they contain bicycles and Roman bricks.

Clearly, many gravels defined as Würm had actually been formed during postglacial time.

By 1969 it was embarrassingly clear that the entire climatic scheme developed for the Alpine terraces by Penck and Brückner, expanded by Eberl, and accepted by a generation of geologists was no more than a house built—not on sand—but on shifting gravel. And when the house finally collapsed, the argument used by Köppen and Wegener to confirm the Milankovitch theory collapsed with it.

As Kukla chipped away at the Penck-Brückner scheme in Europe, Jan van Donk at Lamont was completing isotopic measurements of forams in Caribbean core V12–122 (Figure 38). Along with Broecker, van Donk was attempting to improve the geological time scale. Because the core did not extend to the base of the Brunhes Epoch, the magnetic time scale could not be applied directly. However, the core did contain the U-V boundary—which Ericson had dated as about 400,000 years old by interpolation in cores long enough to contain the last magnetic reversal. This estimate, falling as it did in the middle of the rather

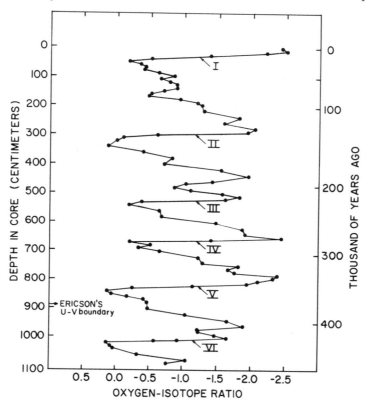

Figure 38. The 100,000-year pulse of climate. Climatic variations shown here are recorded by changes of the oxygen-isotope ratio in a deep-sea core from the Caribbean (V12–122). After determining the approximate time scale, W.S. Broecker and J. van Donk concluded that the major pulsebeat of climate was a 100,000-year cycle. Six intervals of rapid deglaciation are indicated by Roman numerals and referred to as *terminations*. (Adapted from W.S. Broecker and J. van Donk, 1970.)

broad range of dates obtained by uranium and thorium methods, became the cornerstone of Broecker and van Donk's chronology, and led them to conclude that the major cycle in the isotopic record was 100,000 years. Moreover, they noted that this primary climatic cycle had an asymmetrical, sawtoothed shape: "Periods of glacial expansion averaging about 100,000 years in length were abruptly terminated by rapid deglaciations." They labeled these episodes of rapid warming, "terminations."

Not until September 1969, when Broecker and Kukla met at

the international scientific congress in Paris, did they realize that
their separate lines of research had led them to many of the same
conclusions: the major ice ages of the Pleistocene were spaced
about 100,000 years apart, developed slowly, and terminated
abruptly. The marklines in the Czechoslovakian brickyards corresponded to the terminations in the Caribbean cores.

While Broecker and Kukla discussed the shape of Pleistocene
cycles—and Imbrie and Shackleton exchanged views on Pleistocene temperatures—William Ruddiman and Andrew McIntyre
were hard at work at the Lamont Observatory, developing a new
method for studying the history of the ocean. By selecting cores
along a north-south line, and recording the changing distribu-
tions of temperature-sensitive species along that line, they were
able to trace the shifting course of the Gulf Stream. During
interglacial intervals, the current had flowed northeast across the
Atlantic from Cape Hatteras towards Great Britain. But during
the ice ages, it had taken an easterly course towards Spain. As the
ice sheets expanded and contracted, and forests and prairies
swept back and forth across Europe and Asia, the Gulf Stream
swung back and forth like a gate hinged on Cape Hatteras. By
counting the number of "swings" of the current, and keying these
into the magnetic time scale, Ruddiman and McIntyre found that
there were eight climatic cycles within the Brunhes Epoch. Like
the Arctic ice sheets and the Eurasian forests, the ocean's currents
marched to a 100,000-year beat.

By the early 1970s, the reality and importance of the 100,000-
year climatic cycle was obvious. What caused that cycle to occur,
however, was still unclear. The Milankovitch theory itself did not
predict it. Instead, the dominant periodicity of the radiation
curve for the summer season at 65° North latitude was that of the
tilt cycle: 41,000 years.

Nevertheless, Kenneth Mesolella at Brown University and
George Kukla in Czechoslovakia found ways of modifying the
Milankovitch theory to account for the 100,000-year cycle. Both
Kukla and Mesolella believed that changes in orbital eccentricity
were indirectly responsible for the 100,000-year cycle, and they
pointed out that the dominant cycle in the eccentricity curves
(close to 100,000 years) was a good match for the main climatic
pulse. Arguing in much the same way that James Croll had a
century earlier, Kukla and Mesolella emphasized that the intensity of radiation during any particular season is largely controlled
by the precession cycle—the amplitude of which is exactly pro-

portional to eccentricity (Figure 35). When the orbit is unusually elongate, the contrast between seasons is correspondingly great—winters are colder than average and summers are warmer. Therefore, if the temperature during one particular season is critical to the expansion or retraction of ice sheets, it follows that the 100,000-year cycle must be reflected in the climatic record.

But at this point the two theorists parted company. Mesolella believed, as Milankovitch had, that summer was the critical season. Kukla, on the other hand, believed that changes in the amount of winter radiation received at high northern latitudes triggered the ice ages. In a forceful statement published in 1967, Kukla predicted that: "When this problem has been clarified, and the importance of winter accepted, the chance selection of summer will probably be considered the most serious mistake in Quaternary research in recent years."

For their part, Broecker and van Donk were unwilling to commit themselves on the question of the origin of the 100,000-year cycle. Although the four youngest ice-age terminations coincided with inflections of the eccentricity curve, the two oldest terminations did not.

By 1969, the magnetic time scale had proved its value as a basis for studying the history of the ice ages, and had made it possible to identify the dominant pulsebeat of climate as the 100,000-year cycle. But the advent of that time scale had so far done little to generate much support for the astronomical theory. On the contrary, it was something of an embarrassment that the 100,000-year cycle had not been predicted by that theory. Not until after the facts were in had Kukla and Mesolella suggested how the Milankovitch theory could be modified to account for that cycle.

Most scientists, therefore, would only be convinced that the astronomical theory was correct if it could be shown that the small oscillations superimposed on the 100,000-year cycle were those that Milankovitch had predicted. If these shorter climatic cycles turned out to correspond to the 41,000-year cycle of axial tilt, and to the 22,000-year cycle of precession, then the astronomical theory of the ice ages would be confirmed. To demonstrate such a correspondence, however, parallelisms between astronomical and climatic curves must be demonstrated in records sufficiently detailed to exhibit the 22,000- and 41,000-year cycles. Once more, the problem of testing the astronomical theory of the ice ages hinged on increasing the accuracy of the geological time scale.

15

Pacemaker
of the Ice Ages

In the spring of 1970, James D. Hays decided that the time had come to launch a new attack on the ice-age problem. With the Pleistocene time scale fixed at several points by magnetic reversals, and with paleontological techniques available to track ocean currents and to estimate ocean temperatures, deep-sea cores had been transformed into instruments for monitoring global climate. For the first time geologists could determine when—and to what extent—different parts of the ocean changed during ice ages. If a more accurate time scale could be established within the Brunhes Epoch, it would also be possible to make a definitive test of the Milankovitch theory.

But five years of experience with cores from the Antarctic and Pacific Oceans had convinced Hays that a satisfactory reconstruction of oceanic history was too big a task for any single investigator, or any single institution. A team of paleontologists, mineralogists, geochemists, and geophysicists would have to be formed. Discussing this idea with John Imbrie at a lunch counter near Columbia University, Hays pointed out that the appropriate techniques were already being applied by investigators working in a dozen laboratories. All that was needed was an organization to coordinate these independent efforts.

Eager to see the multiple-factor technique applied to species other than forams, Imbrie agreed to participate in the project. He pointed out that Andrew McIntyre and others were already using forams and coccoliths (minute surface-dwelling plants) to map parts of the ice-age Atlantic. If the multiple-factor technique worked with radiolaria and diatoms, then it would be possible to extend McIntyre's results into higher latitudes, and chart the entire ocean. But Imbrie expressed concern that such an interinstitutional effort might be unmanageable. "Don't worry,"

Hays replied, "all we need is money for plane fares and telephone bills."

Hays's optimism was justified. On May 1, 1971 the interdisciplinary, interinstitutional project he had envisioned was under way. Named CLIMAP, the project's first objective was to reconstruct the history of the North Pacific and North Atlantic Oceans during the Brunhes Epoch. Financial support came from the National Science Foundation's International Decade of Ocean Exploration program (IDOE). In 1973, the project was expanded and given two broad goals: to map the surface of the earth during the last ice age, and to measure the oscillations of Pleistocene climate.

Initially, three institutions were represented in the IDOE Project: Lamont-Doherty Geological Observatory of Columbia University, Brown University, and Oregon State University. The Executive Committee consisted of James D. Hays, John Imbrie, Andrew McIntyre, Ted C. Moore, Jr., and Neil Opdyke. Later, the University of Maine and Princeton University joined the project, and the Executive Committee was expanded to include George Denton, Ross Heath, Warren Prell, and William Hutson. George Kukla, now a member of the Lamont staff, assumed responsibility for correlating marine and nonmarine records of climate; Nicholas Shackleton and Jan van Donk measured oxygen-isotope ratios; and Robley K. Matthews analyzed sea-level history. A central administrative office was established at Lamont, and the task of coordinating the far-flung activities was given to Rose Marie Cline. Eventually, nearly a hundred investigators would be involved in the project, including scientists with separate funding in Denmark, France, West Germany, the United Kingdom, Norway, Switzerland, and the Netherlands. In 1976, the group published a global map showing the temperatures of the ocean and the distribution of glaciers at the height of the last ice age, 18,000 years ago. By 1977, expenditures totaled $6,630,500.

But in the spring of 1971, CLIMAP's most urgent task was to subdivide the 700,000-year-long Brunhes Epoch into stratigraphic zones—that is, into layers that could be recognized and correlated from core to core. Only if such a stratigraphic scheme were available would it be possible to recognize erosional gaps, turbidity currents, and other local distortions of the climatic record. Once recognized, these distortions could be avoided or corrected. Ericson had almost solved this stratigraphic problem in 1968, when he subdivided the Brunhes Epoch into ten *menardii* zones (Q through Z). But because these zones were based on the presence or absence of one low-latitude species, Ericson's scheme could not be used with confidence outside of the equatorial Atlantic and Caribbean oceans. What CLIMAP needed was a stratigraphic scheme that could be applied to every ocean.

The task of devising such a scheme was assigned to a core-reconnaissance group that included Tsunemasa Saito, Lloyd Burckle, and Allan Bé. The group was hopeful that Emiliani's oxygen-isotope curve might provide them with the scheme they needed. But because Emiliani's longest core, P6304–9 from the Caribbean, did not penetrate the Brunhes-Matuyama boundary, its 17 isotopic stages floated in a chronologic limbo, somewhere within the Brunhes Epoch.

What Saito and his colleagues needed was a core containing abundant forams and long enough to include the most recent magnetic reversal. In December 1971, Saito located a core (V28–238) which had been raised earlier that year by Lamont scientist John Ladd from shallow waters of the western equatorial Pacific Ocean. The composition of the lowermost foram assemblage convinced Saito that the core was indeed long enough—perhaps it would be the long-sought Rosetta Stone that would enable CLIMAP scientists to decipher the climatic history of the Brunhes Epoch. When Neil Opdyke analyzed the core magnetically, he found that Saito had been right. The Brunhes-Matuyama boundary was 1.2 meters below the top of the core. Realizing the importance of this discovery, Hays immediately sent samples of V28–238 to Nicholas Shackleton at Cambridge University for isotopic analysis.

Hays had met the young British geophysicist several years earlier and had been impressed by the improvements Shackleton had already made in laboratory technique. In 1961, invited by Sir Harry Godwin to join the staff of the Department of Botany at Cambridge, Shackleton had set up a mass-spectrograph for isotopic studies of Pleistocene fossils. Early in his work, he had become convinced that it was important to study isotopic variations in the shells of species that live on the sea floor. But bottom-dwelling species occurred in such low concentrations in the sediment that it was difficult to find enough specimens to make accurate analyses. Shackleton therefore resolved to modify the instrument so that accurate readings could be obtained from a small number of specimens. This modification cost him 10 years.

When Shackleton arrived at the CLIMAP meeting at Lamont in June 1972, he brought with him two isotopic curves for core V28–238. One curve reflected variations in the isotopic composition of planktonic shells formed in near-surface water. That curve, extending down to the most recent magnetic reversal, gave promise of solving CLIMAP's stratigraphic problem—for it showed that the Brunhes Epoch could be divided into nineteen isotopic stages. The top seventeen stages corresponded exactly to those that Emiliani had established in his long Caribbean core. Two additional stages now extended the sequence back to the base of the Brunhes Epoch (Figure 39).

However, the promised solution to CLIMAP's stratigraphic problem would only be fulfilled if it could be demonstrated that the isotopic fluctuations were globally synchronous. Shackleton's colleagues were therefore delighted to discover that his second curve—which reflected variations in the isotopic composition of foraminifera living on the sea floor—was identical to that of the planktonic curve. And as Shackleton pointed out, oceanic bottom water—always close to freezing—could not have been much colder during an ice age. Therefore, as he and Imbrie had suspected since their meeting in Paris three years before, both curves reflected changes in the proportion of light isotopes in the ocean—not changes in water temperature. And, because sea water was mixed rapidly by currents, any chemical change in one part of the ocean would be reflected everywhere within a thousand years. All along, Emiliani's curve had been a chemical message from the ancient ice sheets. When the glaciers expanded, light atoms of oxygen were extracted from the sea and stored in the ice sheets—altering the isotopic ratio of oxygen in sea water. When the glaciers melted, the stored isotopes flooded back into the ocean, returning it to its original composition. The effect of local variation in temperature was too small to be detected.

Shackleton and Opdyke's results not only solved the stratigraphic problem, but also gave CLIMAP investigators a much more accurate chronology for late Pleistocene events. Since the chronology of Shackleton's sequence of 19 isotopic stages was now firmly fixed at both ends—by radiocarbon dates at the top and by a magnetic reversal at the bottom—the age of each stage could now be estimated by interpolation within the 700,000-year Brunhes Epoch.

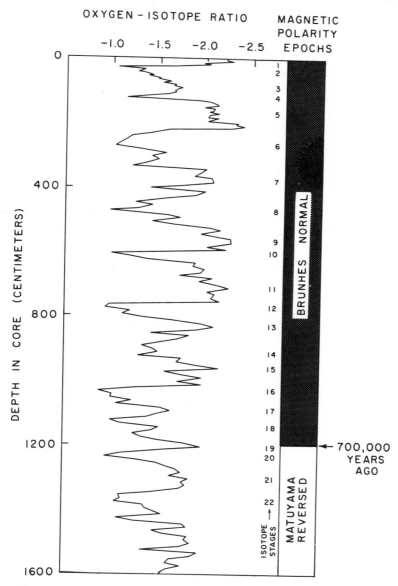

Figure 39. The "Rosetta Stone" of late Pleistocene climate. A graph of the isotopic and magnetic measurements made in 1972 by N.J. Shackleton and N.D. Opdyke on a Pacific deep-sea core (V28–238). These observations—which established that isotope Stage 19 occurs at the boundary between the Brunhes and Matuyama Epochs—provided the first accurate chronology of late Pleistocene climate. (Data from N.J. Shackleton and N.D. Opdyke, 1973.)

With the time scale of the isotopic curve settled at last, Shackleton decided to see if the smaller fluctuations in the isotopic curve corresponded with those predicted by Milankovitch. If the astronomical theory of climate was correct, these fluctuations would reflect variations in axial tilt and precession—and would appear as 41,000-year and 22,000-year cycles superimposed on the main, 100,000-year pulse. But the 100,000-year cycle was so dominant that Shackleton found it difficult to determine what these modulating frequencies were.

Six years earlier, in 1966, a Dutch investigator named E. P. J. van den Heuvel had solved this problem statistically. Using a technique called spectral analysis, he had demonstrated that Emiliani's isotopic curve contained two frequency components: a dominant 40,000-year cycle, and a less distinct 13,000-year cycle. The procedure was similar to that used by a musician analyzing a musical chord into its individual notes. After analyzing the isotopic "chord" into a large number of "notes"—each representing a particular oscillation frequency—he had recorded the relative importance of each oscillation frequency on a graph, or spectrum. The 40,000-year cycle appeared as a distinct peak on the graph—apparently indicating that this cycle was the dominant climatic pulse.

Discussing the approach with Shackleton, Imbrie pointed out that although the spectral method was ideal for testing the Milankovitch theory, the actual results obtained by van den Heuvel must have been biased by his use of a now-abandoned chronology. If the isotope curve were reanalyzed—this time using the CLIMAP time scale—the dominant climatic pulse should be the 100,000-year cycle. What the shorter cycle reported by van den Heuvel would be, Imbrie and Shackleton were not sure; but they resolved to find out by carrying out a spectral analysis of the isotopic curve for core V28–238.

Having already experimented with spectral techniques, Imbrie had the necessary computer programs available at Brown University, and there the two investigators carried out their first statistical experiment. Their results offered some support for the Milankovitch theory. In addition to the 100,000-year cycle which, as expected, appeared as a dominant peak in the calculated spectrum, they found two smaller peaks indicating the presence of climatic cycles about 40,000 and 20,000 years long. Although the amplitudes of these cycles were too small to rule out chance entirely, the near coincidence of the two measured frequencies with the predicted frequencies of tilt and precession was at least suggestive.

Suggestive but not convincing. Why was it proving so difficult to find out what the higher frequencies in the climatic curves were? Reviewing this problem in the fall of 1972, Hays thought he knew the reason: the cores that had so far been analyzed spectrally had accumulated too slowly. He argued that when accumulation rates were as low as one or two millimeters per century—as they were in most Pacific and Caribbean cores—the burrowing activities of animals living on the sea floor would blur the record of the higher frequency cycles. To make a valid test of the Milankovitch theory, therefore, it would be necessary to analyze an undisturbed core whose accumulation rate had exceeded two millimeters per century.

Hays and his CLIMAP colleagues were already studying all of the available cores as part of their effort to map the ice-age ocean. After some reflection, Hays decided that they would now search for a particular type of core: one that had a suitably high sedimentation rate—that was located in the high latitudes of the southern hemisphere—and that contained shells of both forams and radiolaria. Such a core, Hays reasoned, would provide more information than one located in the northern hemisphere. Variations in the isotopic composition of foram shells would provide a record of ice-sheet fluctuations in the northern hemisphere—for nearly all of the glacial expansion and contraction that influenced the isotopic composition of the ocean took place there. At the same time, changes in radiolarian populations could be analyzed by the multiple-factor technique and made to reveal what the history of water temperature had been over the coring site. By comparing the two signals—isotopic and radiolarian—Hays hoped to be able to answer a question that had first been raised by James Croll: do climatic changes in the southern hemisphere coincide with those in the northern hemisphere?

In January 1973, Hays located a core in the Lamont collection that seemed to meet his requirements. Core RC11–120 had been raised from the southern Indian Ocean six years earlier by Geoffrey Dickson aboard the *Robert Conrad*. After counting the radiolaria and sending samples to Shackleton for isotopic analysis, Hays was gratified to find that the deposition rate was high enough for his purposes (three millimeters per century).

When the data were plotted, the answer to Croll's question was immediately apparent: climatic changes in the northern hemisphere were essentially synchronous with those of the southern hemisphere. Although this result alone was important enough to justify his efforts, Hays was disappointed to find that the core only extended back about 300,000 years, to the base of Stage 9 in Emiliani's isotopic scheme. To provide a suitable record for spectral analysis, a core extending back at least 400,000 years would be needed.

When it became clear that the needle Hays was looking for was not to be found in the Lamont haystack, he decided to search elsewhere. In July, he went to Florida State University in Tallahassee, where an extensive collection of Antarctic cores was maintained. There, he continued the search for cores taken near the site of RC11–120. Soon he came upon several cores taken by Norman Watkins aboard the *Eltanin* in 1971. With the assistance of two graduate students, Hays began to open the Watkins cores. Later he would recall: "The cores were kept in cold storage, and we were all shivering in our parkas. But when core E49–18 was opened, we stopped shivering. I knew right away we had something interesting because the color-banding matched perfectly with the oscillations in Shackleton's oxygen curve for V23–238." Counting down, Hays found that the core extended to Stage 13—giving it an age of 450,000 years. He had found his needle at last.

Hays's off-the-cuff stratigraphic analysis turned out to be correct. Core E49–18 did indeed extend back to Stage 13. Unfortunately, the top three isotopic stages had been lost when the core was taken; but with the isotope stratigraphy now available, these could be patched in from the nearby core RC11–120. Together, these two cores contained a detailed and undisturbed record of climate extending back 450,000 years—and their accumulation rate was high enough to have preserved cycles as short as 10,000 years.

When the radiolarian and isotopic data had been graphed, Hays and Shackleton were elated. For the isotope curves in the Indian Ocean matched the general pattern that Emiliani had established for Stages 1 through 13 in a number of other cores. But now, as Hays had anticipated, frequencies higher than the 100,000-year cycle were clearly visible (Figure 40). Realizing that an opportunity to make a definitive test of the Milankovitch

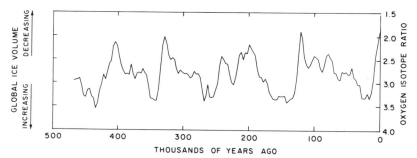

Figure 40. Climate of the past half-million years. A graph of the isotopic measurements made on two Indian Ocean cores by a CLIMAP research group. These observations—which reflect variations in the volume of global ice—led to a confirmation of the astronomical theory of the ice ages. (Data from J.D. Hays et al., 1976.)

theory was at hand, he asked Imbrie to carry out a spectral analysis.

The first objective was to find out exactly what the frequencies of variation in tilt and precession had been over the past 450,000 years (Figure 41). These frequencies, rather than the frequency of the eccentricity cycle, would be crucial to the coming test— because they alone were unambiguously predicted by the Milankovitch theory. Imbrie knew that Anandu D. Vernekar, at the University of Maryland, had recently recalculated the astronomical curves, and Imbrie obtained copies of the calculations from him. After processing Vernekar's information statistically, Imbrie found that, as expected, the tilt curve showed a single cycle of 41,000 years. But the spectrum for the precession curve contained not one, but two distinct cycles—a major precessional cycle of 23,000 years, and a minor cycle of 19,000 years. Concerned that his calculations had been wrong, Imbrie laid his results before Belgian astronomer André Berger. After examining the trigonometrical formulas from which the precession calculations were derived, Berger announced that the double cycle that Imbrie had found was not a statistical error: variations in earth-sun distance do in fact occur as 23,000-year and 19,000-year cycles.

Berger's endorsement set the wheels in motion. According to the expanded version of the astronomical theory developed by Mesolella and Kukla, climatic oscillations should occur as four distinct cycles: a 100,000-year cycle corresponding to variations in eccentricity; a 41,000-year cycle corresponding to variations in

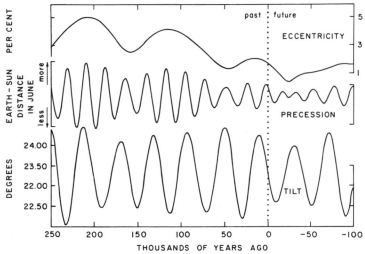

Figure 41. Changes in eccentricity, tilt, and precession. Planetary movements give rise to variations in the gravitational field, which in turn cause changes in the geometry of the earth's orbit. These changes can be calculated for past and future times. (Data from A. Berger.)

axial tilt; and 23,000- and 19,000-year cycles corresponding to variations in precession. In the summer of 1974, Imbrie performed the long-awaited test. Spectral analysis indicated that, as expected, the dominant climatic pulse was a 100,000-year cycle, which appeared on both the isotopic and the radiolarian spectra as a large peak. But three other peaks—smaller but nevertheless distinct—also appeared on the spectra (Figure 42). On the isotopic spectrum these cycles were 43,000 years, 24,000 years, and 19,000 years long. On the temperature-radiolarian spectrum, they were 42,000 years, 23,000 years, and 20,000 years long.

These results were everything for which Imbrie and his colleagues had hoped. Each of the cycles found in the Indian Ocean cores matched the predicted cycles within five percent. That such a coincidence might occur by chance alone seemed highly unlikely. Before long, Nicklas G. Pisias provided additional evidence in support of the astronomical theory. Using a more powerful spectral method, he found a statistically significant 23,000-year cycle in core V28–238. CLIMAP investigators—realizing that their isotope records from the Pacific and Indian oceans matched

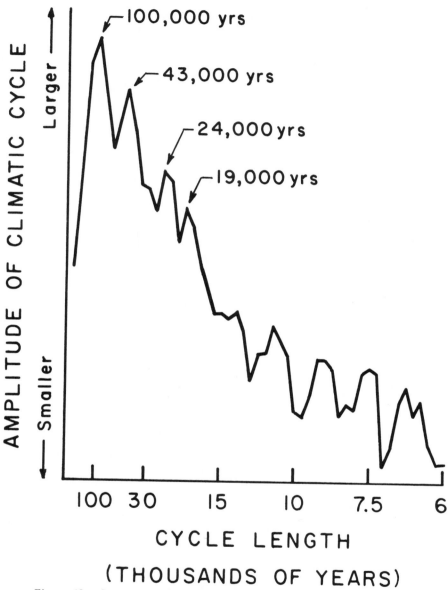

Figure 42. Spectrum of climatic variation over the past half-million years. This graph—showing the relative importance of different climatic cycles in the isotopic record of two Indian Ocean cores—confirmed many predictions of the Milankovitch theory. (Data from J.D. Hays et al., 1976.)

the corresponding parts of isotope records already known from other oceans—felt justified in concluding that the succession of late Pleistocene ice ages had indeed been triggered by changes in the earth's eccentricity, precession, and tilt.

If the astronomical theory were correct, it should be possible to do more than demonstrate by spectral analysis that the astronomical frequencies appeared in climatic curves. It should also be possible to discover how rapidly the ice sheets had responded to each type of astronomical variation. For example, if the ice sheets responded instantly to a change in axial tilt, then the fluctuations of the 41,000-year climatic cycle should have occurred simultaneously with variations in tilt. But if, as seemed more likely, the ice sheets were slow in responding to a change in radiation caused by changes in tilt, the 41,000-year cycle of climate should follow regularly behind the orbital curve.

Discovering that a statistical technique called filter analysis was available to examine the 41,000-year and 23,000-year frequency components of a climatic curve separately, Imbrie applied this method to the records from the two Indian Ocean cores. The result showed clearly that the 41,000-year climatic cycle did lag behind variations in axial tilt by about 8000 years. And, for at least the major portion of the record under study, the 23,000-year climatic cycle lagged systematically behind variations in precession. Moreover, these lags were regular enough to confirm the inference that variations in tilt and precession set the pace for climatic change.

Convinced now that major climatic changes were caused by astronomical variations, and that the 41,000-year and 23,000-year climatic cycles followed systematically behind variations in tilt and precession, Hays, Imbrie, and Shackleton announced their findings in an article in *Science*, which appeared on December 10, 1976: "Variations in the Earth's Orbit: Pacemaker of the Ice Ages."

A century after Croll published his theory and 50 years after Milankovitch mailed his radiation curves to Köppen and Wegener, two cores from the Indian Ocean confirmed the astronomical theory of the ice ages. At last, geologists had clear evidence that the motions of the earth in its orbit around the sun triggered the succession of late Pleistocene ice ages. Exactly how this triggering mechanism operated, and why the 100,000-year cycle of orbital eccentricity appeared to be so strongly impressed on the climatic record of the last half-million years were still unknown. But, for the moment, it was enough to know that Milutin Milankovitch, traveler through distant worlds and times, had led the way to solving a major part of the ice-age mystery.

In March 1941, looking back on a lifetime devoted to finding the cause of the ice ages, Milankovitch had reflected that:

> These causes—the changes in insolation brought about by the mutual perturbations of the planets—lie far beyond the vision of the descriptive natural sciences. It is therefore the task of the exact natural sciences to outline this scheme, by means of its laws ruling the universe and by its developed mathematical tools. It is left, however, to the descriptive natural sciences to establish an agreement between this scheme and geological experience.

PART III

ICE AGES
OF THE FUTURE

16

The Coming Ice Age

What of the future? Does the fact that ice ages have occurred many times in the past mean that another one lies ahead? Unless there is some fundamental and unforeseen change in the climate system, most scientists who have examined the evidence agree that the world will experience another age of ice. But when? On this question, geologists disagree. Some predict that the present interglacial age will last another 50,000 years. Others, finding that the earth has been cooling for some time, believe that an ice age is already on its way—due within the next few centuries, according to one extreme view.

To some extent, this clash of opinion is only a matter of words. What defines the beginning of an ice age? How extensive must an ice sheet be, and how far must global temperatures fall before the world is "officially" declared to be in an ice age? The problem of definition is complicated by the facts of geography. Much of Greenland, for example, is presently experiencing an ice age. If the Greenland Ice Sheet were to expand even one percent, so that houses along the coast were destroyed, families living in those houses might well conclude that an ice age was beginning. But at the same time, a Scottish fisherman—used to foul weather even in the best of times and far-removed from the ice sheets slowly expanding in Greenland and Scandinavia—might be unaware that an expansion had occurred. Only later, when ice caps appeared on the summit of Ben Nevis and the schools of herring migrated south, would the Scottish fisherman be likely to judge that the interglacial age had ended. And many more thousands of years would pass before the farmlands of Central Europe were replaced by a polar desert and the rain forests of Brazil gave way to grassland.

The problem of defining an ice age has been settled somewhat arbitrarily on the basis of Pleistocene deposits in Central Europe.

Here, past interglacials are easily placed within time boundaries. They begin abruptly, with the arrival of broad-leafed trees and forests, and end when deciduous forests disappear and are replaced by grassland. By common agreement, therefore, a Pleistocene interglacial age is defined as any interval of time during which oak and other deciduous trees are widespread in Europe. It is the demise of these oak forests that signals the beginning of an ice age.

On the basis of this definition, the present interglacial age—the Holocene Epoch—began about 10,000 years ago. The problem of predicting when it will end can be approached in several different ways, some of which are more satisfactory than others. One approach treats the geological record of climate statistically, and uses the known duration of previous interglacials as a basis for estimating the remaining lifespan of the present interglacial—in much the same way that a life insurance company calculates the probable lifespan of an individual. An analysis of deep-sea cores (Figure 40), shows that no Pleistocene interglacial has lasted more than about 12,000 years and that most have had lifespans of about 10,000 years. Statistically speaking, then, the present interglacial is already on its last legs, tottering along at the advanced age of 10,000, and can be expected to end within the next 2000 years.

But only insurance companies are satisfied by statistical predictions of this kind. A more satisfactory way of predicting the end of the present interglacial age is to project current climatic trends into the future. One such trend is the long-term cooling that began 7000 years ago, at a time known as the Postglacial Climatic Optimum, when temperatures were warmer and rainfall was greater than today. Since then, the average temperature has been gradually declining (Figure 43). As discussed further below, short episodes of warming and cooling—known as the Little Ice Age Cycle—have been superimposed on this general cooling trend. The net result has been a 2°C lowering of the average global temperature. The clearest indications of this trend are changes in the geographic ranges of animals and plants. For example, species of oak trees and edible mussels that are today entirely absent from Scandinavia flourished there 7000 years ago. Elsewhere in Europe, vegetation belts have either moved steadily southward, or have been driven to lower elevations. If this trend were to continue, global temperatures would reach ice-age levels (6°C cooler than they are today) some 18,000 years from now.

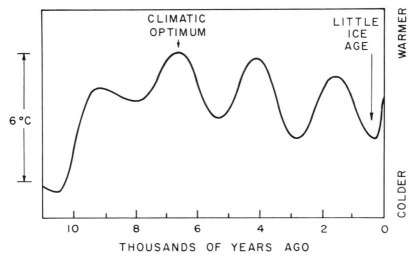

CLIMATIC OPTIMUM

LITTLE ICE AGE

WARMER

COLDER

6 °C

10   8   6   4   2   O

THOUSANDS OF YEARS AGO

Figure 43. Climate of the past 10,000 years. This graph shows general trends in global temperature, as estimated from geological records of mountain glaciers and fossil plants. During the Climatic Optimum, temperatures were about 2° C warmer than they are today. About 300 years ago, during a climatic episode known as the Little Ice Age, temperatures were cooler than they are today.

Exactly how much cultural impact the recorded 2° C drop in average temperature has had, is difficult to assess. But there is no doubt at all that the decrease in rainfall—which in most places has accompanied the cooling—has had a definite effect on patterns of agricultural production, and therefore on patterns of human settlement. Michael Sarnthein, a German geologist who has made a worldwide study of the evidence, concludes that the total area covered by sandy deserts has significantly increased since the Climatic Optimum. For example, regions in North Africa that are dry and unproductive today, had sufficient rainfall during the Climatic Optimum to support great civilizations.

A cooling trend of much shorter duration than 7000 years was first identified by J. Murray Mitchell, Jr. in 1963. By averaging thermometer readings made at a worldwide network of weather stations, Mitchell was able to show that global climate has been cooling since 1940 (Figure 44). Over a 20-year period, he found, the average temperature of the northern hemisphere has fallen about 0.3° C. If this trend were to continue, average temperatures at many places would reach ice-age levels only 700 years from

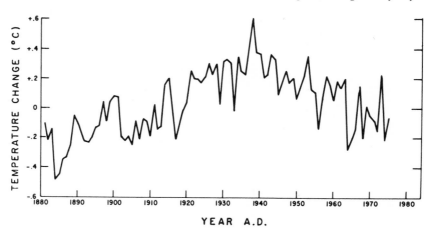

YEAR A.D.

Figure 44.   Climate of the past 100 years. This graph shows changes in the average annual temperature of the northern hemisphere. Since 1939, average temperatures have declined about 0.6°C. (From J.M. Mitchell, Jr., 1977.)

now. Long before then, changes in rainfall would undoubtedly disrupt existing patterns of food production and bring about drastic cultural changes. As Mitchell pointed out, however, there is only one thing certain about climatic trends: they can and do reverse themselves. And by the mid-1970s, it was by no means clear that the cooling trend that began in 1940 was continuing.

The fallibility of climatic predictions based on short-term trends is well demonstrated by the annual march of the seasons. Yet some primitive peoples—failing to recognize the cyclic nature of the seasonal succession—became alarmed each year when they perceived a four-month cooling trend, and built bonfires to encourage the sun to return to his duties. This example illustrates that what an observer perceives as a trend may only be one phase of a cycle. Forecasts based on observed trends are therefore valid only if the cycles that drive them are understood. And since no one has yet been able to produce a convincing explanation of Mitchell's trend, those prophets of doom who use such data to predict an early end of the present interglacial age are in danger of repeating the error made by the builders of bonfires many centuries ago.

The astronomical theory of the ice ages provides a basis for forecasting the course of future climate that avoids the uncertainties inherent in predictions based on trends. As shown in Figure

41, the problem of making such a forecast is complicated by the fact that changes in eccentricity and tilt are now working to cool the climate—while the precession cycle is working to warm the climate. How will these effects combine? To find out, John Imbrie and John Z. Imbrie have developed a mathematical formula for estimating the volume of global ice directly from André Berger's astronomical curves. When their formula is applied, the result indicates that the cooling trend that began 7000 years ago will continue into the future, and lead to a maximum advance of the glaciers 23,000 years from now.

However, the long-term cooling trend predicted by the astronomical theory will undoubtedly be modified by climatic oscillations of much shorter duration than the precession cycle. Such oscillations have already occurred many times during the Holocene, and there is every reason to expect that similar events will occur again. The best-known of these small oscillations is the Little Ice Age, which lasted from about A.D. 1450 to 1850 (Figure 45). During that 400-year interval, valley glaciers in the Alps, Alaska, New Zealand, and Swedish Lapland advanced well beyond their present limits and snow lay for months on the high mountains of Ethiopia, where it is now unknown (Figure 46).

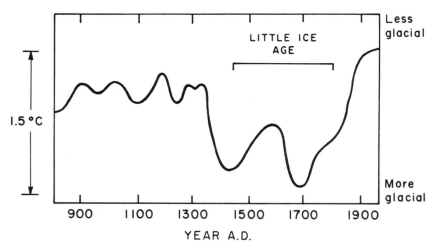

Figure 45. Climate of the past 1,000 years. The graph is an estimate of winter conditions in Eastern Europe, as compiled from manuscript records. During the Little Ice Age (A.D. 1450–1850), mountain glaciers all over the world advanced considerably beyond their present limits. (Adapted from H.H. Lamb, 1969.)

Figure 46. The Argentière glacier today and in 1850. Top: In a photograph taken in 1966, the glacier is seen as a small tongue of ice in the upper portion of the valley. Bottom: An etching made about 1850 shows the extent of the glacier during the waning phase of the Little Ice Age in the French Alps. (From L. Ladurie, 1971.)

Global climate generally was 1° C cooler than now. According to Hubert H. Lamb, who delved into old manuscripts and logbooks to piece together the story of the Little Ice Age, the picture of Hans Brinker skating on Dutch canals is an accurate reflection of the severe winters that characterized European climate during the Little Ice Age.

Meanwhile, colonists in New England endured winters far more severe than any today. According to David M. Ludlum, who has made a study of climate in colonial times, the legendary winter during which Washington's troops were bivouacked at Valley Forge was actually regarded by contemporary observers as "notably mild." In fact, if Washington had camped at Valley Forge two years later, during the winter of 1779–80, the suffering of his troops would have been much greater. Even by Little Ice Age standards, that winter was "The most hard difficult winter . . . that ever was known by any person living." To the north, New York harbor was frozen solid. Ludlum writes:

> Though both the Hudson and the East Rivers were accustomed to freeze solidly from time to time in the olden days, there was no record of the entire Upper Bay congealing for a number of days . . . . [In late January] people walked on the ice all the way from Staten Island to Manhattan Island, a distance of five miles . . . . Heavy loads and even large cannon were dragged across the iceways to fortify the British position on Staten Island which had been subject to cross-the-ice forays from Washington's outposts in New Jersey.

Through detailed studies of glacial moraines, George Denton and Wibjörn Karlén have shown that the Little Ice Age peaked about A.D. 1700 and was the last of five similar Holocene events: "Viewed as a whole, therefore, the Holocene experienced alternating intervals of glacier expansion and contraction that were probably superimposed on the broad . . . [cooling trend]. Expansion intervals lasted up to 900 years and contraction intervals up to 1750 years." The dates of these glacial maxima—about 250, 2,800, 5,300, 8,000, and 10,500 years ago—suggest that a little-ice-age cycle about 2,500 years long is superimposed on the much longer cycle of major ice ages (Figure 43).

Although the cause of the little-ice-age cycle is unknown, there is some evidence that hints that it is related to variations in the sun. Whatever its cause, the cycle discovered by Denton and Karlén must be taken into account when making a forecast of future climate. Evaluated in terms of the average temperature changes, the cycle of little ice ages has about one-tenth the impact of the cycle of great ice ages. But changes in the higher frequency cycle occur much more rapidly than those due to orbital changes. If Denton and Karlén are correct, the warming effect of the present cycle (which began to be felt about the year 1700) will soon override the cooling effect of the astronomical cycle, and cause temperatures to rise for the next 1000 years. At that time, the astronomically driven cycle and the little-ice-age cycle will combine forces and initiate a long cooling trend that will reach a climax 23,000 years from now.

Such is the forecast suggested by present knowledge of *natural* climatic cycles: 1000 years of warming followed by 22,000 years of cooling. But this forecast does not take into consideration the impact of an "unnatural" agent. J. Murray Mitchell, Jr. writes:

> Should nature be left to her own devices, without interference of man, I feel confident in predicting that future climate would alternately warm and cool many times before shifting with any real authority toward the next ice age . . . . Because of man's presence on the Earth, however, what will actually happen in future decades and centuries may well follow a different scenario; imperceptibly different at first, but importantly so later on. It seems likely that industrial man already has started to have an impact on global climate, although this is difficult to prove by direct observation . . . . If man continues his ever-growing consumption of energy, however, and in the process adds further pollution to the global atmosphere, it may not be very many more years or decades before his impact will break through the noise level of natural climatic variability and become clearly recognizable.

Although many human activities influence climate (for example, agriculture, irrigation, forest cutting, urbanization, and accompanying discharges of heat and smoke), by far the greatest impact on climate will come from the burning of fossil fuels and the accompanying production of carbon dioxide gas. This pollutant is the inevitable product of combustion of all hydrocarbon fuels, including coal, oil, gasoline, natural gas, methane, propane, and a variety of lesser fuels. Since atmospheric carbon dioxide acts as a thermal blanket, the inevitable result of burning fossil fuels will be a worldwide rise in average temperature (Figure 47).

"If our society continues to rely for a very long time on fossil fuels to meet its energy needs," Mitchell writes, "the consequences to climate are likely to become noticeable by the end of this century, but will not become a serious problem until well into the next century."

Viewed in a geological perspective, it is likely that consumption of the bulk of the world's known fossil fuel reserves would plunge the planet into a "super-interglacial age," unlike anything experienced in the last million years. Moreover, the effect of carbon dioxide would endure for a thousand years or more after the use of fossil fuels ceased, for it would take that long for the atmosphere to rid itself of the excess carbon dioxide. Although the exact consequences of such a super-interglacial are difficult to estimate, Mitchell concludes that: "A thousand years of unusually

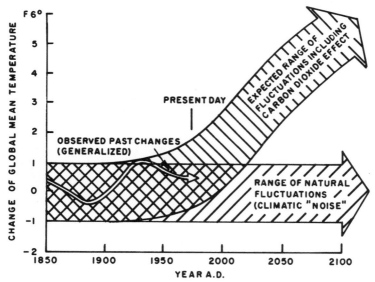

Figure 47. Climatic forecast to the year A.D. 2100. Since 1850, the average global temperature (shown as a white line) has fluctuated through a range of about 2° F. Over the next two centuries, it is likely that temperatures will rise, owing to an expected increase in atmospheric carbon dioxide. This effect may not be evident until about the year 2000. Thereafter, however, the warming may be dramatic. (From J.M. Mitchell, Jr., 1977.)

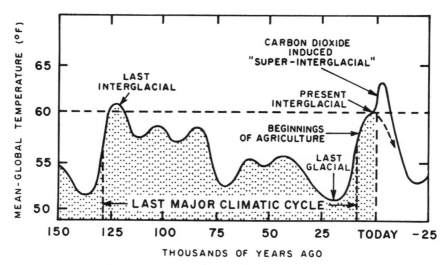

Figure 48. Climatic forecast of the next 25,000 years. According to the astronomical theory of the ice ages, the natural course of future climate (shown by the dashed line) would be a cooling trend leading to full glacial conditions 23,000 years from now. The warming effect of carbon dioxide, however, may well interpose a "super-interglacial," with global mean temperatures reaching levels several degrees higher than those experienced at any time in the last million years. In that case, onset of a cooling trend leading to the next ice age would be delayed until the warming had run its course, perhaps 2,000 years from now. (Modified from J.M. Mitchell, Jr., 1977.)

warm climate would be likely to result in substantial melting of the Greenland and Antarctic ice caps, raising sea levels around the world enough to submerge many of our coastal population centers and much productive farmland." But in some regions, a warmer climate would bring distinct benefits. For example, deserts in North Africa and the Middle East might bloom again, as they did 7000 years ago during the Climatic Optimum.

Assuming that the shock of the super-interglacial does not bring about a fundamental change in the earth's climate system, the atmosphere would eventually rid itself of the excess carbon dioxide. Then the long-term cooling cycles—driven by changes in the earth's orbit and by the cooling phase of the little ice age cycle—would reassert themselves (Figure 48). Some 2000 years from now, a distinct cooling trend would begin. After approximately another 1000 years, the North African deserts would become dry once more, oak forests would disappear from Central Europe, and the longest Pleistocene interglacial on record would come to an end. Global climate would then start a long downward slide until, 23,000 years from now, the earth would once more find itself in the depths of a new ice age.

The Last Billion Years
of Climate

This book has dealt with the climatic history of the last half-million years. During that interval, the great Pleistocene ice sheets underwent periodic expansions and contractions as the earth entered and left a succession of ice ages and interglacial ages. Although the Pleistocene interglacial ages were relatively warm, substantial bodies of permanent ice existed in the polar regions during even the warmest of these climatic regimes—on the continent of Antarctica, on Greenland, and in the surface waters of the Arctic Ocean.

If we look back over a much longer period of time, we discover that only three times during the last billion years have the polar regions been covered with ice sheets comparable in size to those found during the last half-million years. In Figure 49, these long intervals are identified as glacial ages. During each glacial age, ice accumulated at the poles, and the continents were repeatedly glaciated. The first of these periods of repeated glaciation occurred during late Precambrian time, about 700 million years ago. The second interval of glaciation, known as the Permo-Carboniferous Glacial Age, occurred about 300 million years ago. The Present (or Late Cenozoic) Glacial Age began about 10 million years ago.

Exactly why the earth entered and left these long regimes of repeated continental glaciation is not clear, but a reasonable case can be made that these changes in climate were caused by continental drift—the process that results in a slow but continuous change in the geographic position of the continents. According to this theory, an unstable mass of ice accumulates in high latitudes whenever a substantial portion of the earth's land area is located near the poles. In general, the facts about Permo-Carboniferous glaciation fit this theory, for in those remote times the earth's land

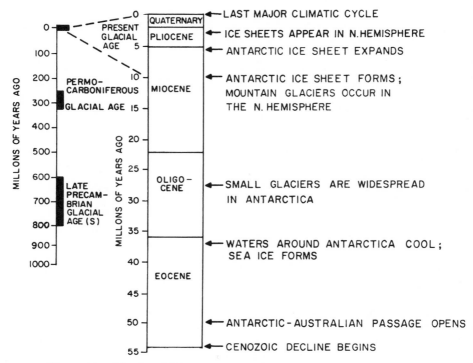

Figure 49. The last billion years of climate. Intervals when ice sheets occurred in polar regions are indicated on the left as glacial ages. An outline of significant events in the Cenozoic climate decline is given on the right.

masses were assembled in a single supercontinent, known as Pangaea. Although Pangaea was centered on the equator, its southernmost tip included the South Pole. The glaciated areas—Brazil, Argentina, South Africa, India, Antarctica, and Australia—were located then in high southern latitudes.

During the 200 million years that followed the Permo-Carboniferous Glacial Age, the earth returned to a nonglacial regime, and was often considerably warmer than today. This condition apparently resulted from a northward movement of Pangaea so that its southern tip no longer included the South Pole. Then, about 55 million years ago, global climate began a long cooling trend that has continued until the present time. Known as the Cenozoic climate decline, this trend is associated with a gradual breaking-up of Pangaea into the separate continents we know today. Antarctica separated from Australia and gradually shifted southward into its present position centered over the South Pole. At the same time, the continents of North America and Eurasia moved toward the North Pole. As more and more land became concentrated in the high latitudes of both polar hemispheres, surface reflectivity increased and climate cooled. About 10 million years ago, small mountain glaciers appeared in Alaska and elsewhere at high latitudes of the northern hemisphere. But the new climatic regime was felt even more dramatically in the southern hemisphere, where the Antarctic Ice Sheet, expanding rapidly to about half of its present volume, became a permanent feature of the present glacial age. About 5 million years ago, the Antarctic Ice Sheet expanded again, and may have been even larger than it is today.

Three million years ago, continental ice sheets appeared for the first time in the northern hemisphere, where they occupied areas adjacent to the North Atlantic Ocean. Apparently, once the ice sheets in this hemisphere formed, they were sensitive to astronomical variations—and they began a long and complex series of fluctuations. So far, it has not been possible to analyze the early history of these fluctuations in detail. But cycles of 100,000 years, 41,000 years, and about 22,000 years are clearly stamped on the climatic record of the most recent half-million years. It is these cycles that are explained by the astronomical theory of the ice ages.

APPENDIX

Chronology of Discovery

1815   Jean-Pierre Perraudin, a Swiss mountaineer from the Val de Bagnes, becomes convinced that Alpine glaciers formerly extended well beyond their present limits.

1818   Ignace Venetz, a highway engineer working in the Val de Bagnes, meets Perraudin and is convinced by his arguments that some Alpine glaciers formerly extended at least five kilometers beyond their present limits.

1836   On field excursions in the Alps, Jean de Charpentier and Ignace Venetz convince Louis Agassiz that many features of lowland landscapes must have been formed by glaciers.

1837   Louis Agassiz announces his theory of a Great Ice Age at a meeting of the Swiss Society of Natural Sciences in Neuchâtel.

1839   Timothy Conrad uses Agassiz' glacial theory to explain surficial deposits of sediment in the United States.

1840   Louis Agassiz persuades William Buckland that surficial deposits in Britain are of glacial origin. Shortly thereafter, Buckland convinces Charles Lyell.

1841   In Scotland, Charles Maclaren argues that the sea surface during an ice age must have been 800 feet lower than its present level.

1842   In France, Joseph Adhémar proposes an astronomical theory of the ice ages based on the precession of the equinoxes.

1843   The French astronomer Urbain Leverrier develops formulas for calculating past changes in the earth's orbit, and reconstructs the orbital history of the past 100,000 years.

1863   Archibald Geikie compiles sufficient field evidence to convince most geologists that surficial deposits in Scotland are of glacial origin.

1864   In Scotland, James Croll publishes an astronomical theory of the ice ages based on the precession of the equinoxes and on changes in orbital eccentricity.

1865   Using records of ancient Scottish shorelines, Thomas Jamieson argues that the weight of Pleistocene ice sheets was sufficient to depress underlying land masses.

1870   In America, Grove Gilbert demonstrates that Great Salt Lake is a remnant of a far larger lake that occupied the region during the last ice age.

        After studying the deserts of Central Asia, Baron Ferdinand von Richthofen concludes that the deposits of yellowish silt (loess), found in nonglaciated regions of Europe, North America, and South America, were deposited during the last ice age by wind.

1871   Amos Worthen demonstrates that more than one ice age occurred in Illinois.

1874   James Geikie, on the staff of the Geological Survey of Scotland, publishes a synthesis of information on the Pleistocene ice ages.

1875   Scientists aboard *H.M.S. Challenger* return from their pioneering oceanographic expedition with extensive information about deep-sea deposits.

1894   James Geikie, now Professor of Geology at the University of Edinburgh, expands his summary of Pleistocene history to include glacial maps of North America, Europe, and Asia.

        Professor James Dana of Yale University rejects Croll's astronomical theory on the grounds that the last American ice sheets disappeared 10,000 rather than 80,000 years ago.

1904   In Germany, Ludwig Pilgrim calculates how the eccentricity, tilt, and precession of the earth's orbit varied over the past million years.

1906   Bernard Brunhes finds evidence in French lava flows that the direction of the earth's magnetic field has changed.

1909   Albrecht Penck and Eduard Brückner use observations on Alpine river terraces to reconstruct the succession of Pleistocene ice ages.

1920   The Yugoslavian mathematician Milutin Milankovitch publishes formulas for calculating the intensity of incoming solar radiation as a function of latitude and season; states that the same calculations could be made for past times; and argues that the climatic effect of changes in radiation would be sufficient to cause ice ages.

1924   In Germany, Wladimir Köppen and Alfred Wegener publish the three curves calculated by Milankovitch, which are the basis of his theory of the ice ages. The curves show how summertime radiation at latitudes 55°, 60°, and 65° North varied as a function of time over the past 650,000 years.

1929   Motonori Matuyama finds evidence in Japan and Korea that the earth's magnetic field reversed itself sometime during the Pleistocene Epoch.

1930   Barthel Eberl elaborates the scheme of Pleistocene history developed by Penck and Brückner and finds that the geological record of Alpine terraces matches the Milankovitch radiation chronology.

1935   Wolfgang Schott discovers a paleontological record of the last ice age in short cores raised from the floor of the equatorial Atlantic Ocean by the German *Meteor* Expedition of 1925–27.

1938   Milutin Milankovitch publishes the final version of his astronomical theory of the ice ages. The main cause is identified as variations in summertime radiation at high latitudes in both hemispheres—variations that result primarily from variations in axial tilt (41,000-year cycle), but which also include the effect of equinoxial precession

(22,000-year cycle). Taking changes in the reflective power of the earth into account, he also calculates how the geographic positions of ice-sheet margins varied over the past million years.

1947   At the University of Chicago, Harold Urey publishes the theoretical basis of the oxygen-isotope method.

In Sweden, Björe Kullenberg invents a piston-coring device used by scientists of the Swedish Deep-Sea Expedition (1947–48) to recover long sections of deep-sea sediment.

1951   Willard Libby develops the radiocarbon dating method at the University of Chicago.

Samuel Epstein and his colleagues at the University of Chicago develop a procedure for calculating ancient ocean temperatures based on Urey's isotope theory.

1952   At the Scripps Institution of Oceanography, Gustaf Arrhenius shows that fluctuations in the chemical composition of Pacific deep-sea cores raised by the Swedish Expedition are records of changing climate.

David Ericson and his colleagues at the Lamont Geological Observatory of Columbia University develop techniques for recognizing turbidity-current layers in deep-sea sediments.

1953   Ingo Schaefer finds fossils in Alpine terrace gravels suggesting that the sequence of glacial and interglacial ages reconstructed by Penck and Brückner is not valid.

Fred Phleger and his colleagues at the Scripps Institution of Oceanography find paleontological evidence of nine ice ages in piston cores of the Atlantic sea floor.

1955   At the University of Chicago, Cesare Emiliani discovers that fluctuations in the oxygen-isotope composition of forams in deep-sea cores record at least seven glacial and seven interglacial stages, and estimates that the main climatic cycle is about 40,000 years long.

1956   John Barnes and his colleagues at the Los Alamos Scientific Laboratory develop the thorium method for dating fossil corals.

David Ericson and Goesta Wollin use changes in the species composition of forams in deep-sea cores to record variations in Pleistocene climate.

1961   George Kukla and Vojen Ložek of the Czechoslovakian Academy of Sciences demonstrate that the sequence of soils and loesses in the nonglaciated regions of central Europe contain detailed records of Pleistocene climate.

1963   Allan Cox and his colleagues at the U.S. Geological Survey demonstrate the synchroneity of magnetic reversals and construct a paleomagnetic time scale.

1964   At the Scripps Institution of Oceanography, Christopher Harrison and Brian Funnell find the Brunhes-Matuyama magnetic reversal in deep-sea cores.

Garniss Curtis, Jack Evernden, and their colleagues at the University of California demonstrate that the potassium-argon method yields reliable ages for Pleistocene events.

1965   At the Lamont Geological Observatory, James Hays uses fossil radiolaria to monitor the Pleistocene history of the Antarctic Ocean.

Wallace Broecker, of the Lamont Observatory, argues that the Milankovitch theory is supported by 80,000-year and 120,000-year thorium dates for interglacial sea-levels.

1966   Cesare Emiliani, now at the University of Miami's Institute of Marine Science, analyzes a long deep-sea core from the Caribbean (P6304-9) that extends his sequence of isotope stages back to Stage 17; and develops a revised time scale that implies that the main climatic cycle is about 50,000 years long.

Robley Matthews and Kenneth Mesolella of Brown University prove that terraces on the island of Barbados were formed by ancient coral reefs. Each reef is therefore a record of an interglacial level of the sea.

1967   At Cambridge, Nicholas Shackleton presents evidence suggesting that variations in oxygen-isotope ratios in deep-sea cores reflect variations in the total volume of the ice sheets.

Geoffrey Dickson, aboard Lamont's *R. V. Robert Conrad*,

raises core RC11–120 from the floor of the southern Indian Ocean.

At the Lamont Geological Observatory, James Hays and Neil Opdyke use magnetic reversals to date climatic events in deep-sea cores from the Antarctic Ocean.

1968   Wallace Broecker, Robley Matthews, and their colleagues at Columbia and Brown Universities report thorium dates from three coral reef terraces on Barbados that correspond to interglacial episodes predicted by a revised version of the Milankovitch theory.

George Kukla and his colleagues at the Czechoslovakian Academy of Sciences use the paleomagnetic time scale to demonstrate that the main climatic fluctuation recorded in European soils is a 100,000-year cycle.

1970   Wallace Broecker and Jan van Donk demonstrate that the main climatic cycle recorded isotopically in Caribbean cores is a 100,000-year cycle.

1971   William Ruddiman, of the U.S. Naval Oceanographic Office, uses the paleomagnetic time scale to show that changes in Atlantic currents correlate with the 100,000-year climatic cycle.

John Ladd, aboard Lamont's *R. V. Vema*, raises core V28–238 from the floor of the western equatorial Pacific Ocean.

James Hays (at the Lamont-Doherty Observatory) and William Berggren (of the Woods Hole Institution of Oceanography) correlate the Pliocene-Pleistocene boundary with the Olduvai Magnetic Event, and thus establish that the Pleistocene Epoch is about 1.8 million years long.

Norman Watkins, aboard the National Science Foundation's *R. V. Eltanin*, raises core E49–18 from the floor of the southern Indian Ocean.

At Brown University, John Imbrie and Nilva Kipp develop a statistical method for estimating the temperature of Pleistocene oceans from a census of microfossil species; and use spectral analysis of faunal and isotopic data from Caribbean core V12–122 in an unsuccessful attempt to find climatic cycles corresponding to variations in tilt and precession.

Members of the National Science Foundation CLIMAP Project begin their effort to extract a global record of Pleistocene climate from deep-sea cores.

1972   Anandu Vernekar, of the University of Maryland, calculates how the geometry of earth's orbit and the intensity of incoming solar radiation vary as a function of time over the past 2 million and next 100,000 years.

Nicholas Shackleton (at Cambridge) and Neil Opdyke (at Columbia) provide a time scale for climatic events of the past 700,000 years by correlating isotopic and magnetic observations on Pacific core V28–238; extend the number of oxygen-isotope stages to 22; and demonstrate that variations in the oxygen-isotope ratio reflect changes in the total volume of the ice sheets.

1975   George Kukla, now at the Lamont-Doherty Geological Observatory, summarizes evidence that demonstrates that the glacial-interglacial succession developed by Penck and Brückner and elaborated by Eberl, is not valid.

1976   By carrying out a spectral analysis of Indian Ocean cores RC11–120 and E49–18, CLIMAP investigators James Hays, John Imbrie, and Nicholas Shackleton establish that major changes in climate have followed variations in earth's tilt and precession over the past 500,000 years—as predicted by the astronomical theory of the ice ages.

SUGGESTED READING

Bryson, R.A. and T.J. Murray, 1977, *Climates of hunger*, Univ. Wisconsin Press, Madison.

Calder, N., 1974, *The weather machine*, British Broadcasting Corp., London.

Eiseley, L., 1958, *Darwin's century*, Doubleday, Garden City, New York.

Fagan, B.M., 1977, *People of the earth*, Little, Brown & Co., Boston.

Gillispie, C.C., 1951, *Genesis and geology: the impact of scientific discoveries upon religious beliefs in the decades before Darwin*, Harper and Brothers, New York.

Ladurie, E.L., 1971, *Times of feast, times of famine: a history of climate since the year 1000*. (Trans. by Barbara Bray). Doubleday, New York.

Lamb, H.H., 1966, *The changing climate: selected papers*, Methuen, London.

Ludlum, D., 1966, *Early American winters: 1604–1820*, American Meteorological Society, Boston.

Lurie, E., 1960, *Louis Agassiz: a life in science*, Univ. Chicago Press, Chicago.

Schneider, S.H., with L.E. Mesirow, 1976, *The genesis strategy*, Dell, New York.

Sparks, B.W. and R.G. West, 1972, *The ice age in Britain*, Methuen, London.

Sullivan, Walter, 1974, *Continents in motion*, McGraw-Hill, New York.

BIBLIOGRAPHY

Adhémar, J.A., 1842, *Révolutions de la mer*, privately published, Paris.

Adie, R.J., 1975, Permo-Carboniferous glaciation of the southern hemisphere, in *Ice ages: ancient and modern*, (A.E. Wright and F. Moseley, eds.), Seel House, Liverpool, pp. 287–300.

Agassiz, L., 1840, *Etudes sur les glaciers*, privately published, Neuchâtel.

Andrews, J.T., 1974, *Glacial isostasy*, Dowden, Hutchinson, and Ross, Stroudsburg.

Angelitch, T.P., 1959, Milutin Milankovitch, *Archives internationales d'histoire des sciences*, *12*, pp. 176–178.

Arrhenius, G., 1952, Sediment cores from the East Pacific, *Swedish Deep-Sea Expedition (1947–1948) Repts.*, *5*, Elander, Göteborg, pp. 1–207.

Barnes, J.W., E.J. Lang, and H.A. Potratz, 1956, Ratio of ionium to uranium in coral limestone, *Science*, *124*, pp. 175–176.

Berger, A., 1977(a), Support for the astronomical theory of climatic change, *Nature, Lond.*, *269*, pp. 44–45.

Berger, A., 1977(b), Long-term variation of the earth's orbital elements, *Celestial Mech.*, *15*, pp. 53–74.

Bernhardi, R., 1832, An hypothesis of extensive glaciation in prehistoric time, in *Source book in geology*, (K.T. Mather and S.L. Mason, eds.), McGraw-Hill, New York, 1939, pp. 327–328.

Bloom, A.L., W.S. Broecker, J.M.A. Chappell, R.K. Matthews, K.J. Mesolella, 1974, Quaternary sea level fluctuations on a tectonic coast, *Quaternary Research*, *4*, pp. 185–205.

Broecker, W.S., 1965, Isotope geochemistry and the Pleistocene climatic record, in *The Quaternary of the United States*, (H.E. Wright, Jr. and D.G. Frey, eds.), Princeton Univ. Press, Princeton, pp. 737–753. This source was extensively used in writing Chapter 12.

Broecker, W.S., 1975, Climatic change: are we on the brink of a pronounced global warming?, *Science, 189*, pp. 460–463.

Broecker, W.S., D.L. Thurber, J. Goddard, T. Ku, R.K. Matthews, and K.J. Mesolella, 1968, Milankovitch hypothesis supported by precise dating of coral reefs and deep-sea sediments, *Science, 159*, pp. 1–4.

Broecker, W.S. and J. van Donk, 1970, Insolation changes, ice volumes, and the O^{18} record in deep-sea cores, *Reviews of Geophysics and Space Physics, 8*, pp. 169–197.

Brunhes, B., 1906, Recherches sur la direction d'aimantation des roches volcaniques, *Jour. de Physique Théorique et Appliquée, Series 4, 5*, pp. 705–724.

Calder, N., 1974, Arithmetic of ice ages, *Nature, Lond., 252*, pp. 216–218.

Carozzi, A.V. (editor), 1967, *Studies on glaciers preceded by the discourse of Neuchâtel by Louis Agassiz*, Hafner, New York. This source was extensively used in Chapter 1.

Charlesworth, J.K., 1957, *The Quaternary Era with special reference to its glaciation, 2 vols.*, Edward Arnold, London. This source was extensively used in writing Chapter 9.

CLIMAP Project Members, 1976, The surface of the ice-age earth, *Science, 191*, pp. 1131–1144.

Collomb, E., 1847, *Preuves de l'existence d'anciens glaciers dans les vallées des Vosges*, Victor Masson, Paris.

Committee for the study of the Plio-Pleistocene boundary, 1948, *Int. Geol. Congr. Rep. 18th Session, Great Britain, 9*.

Conrad, T.A., 1839, Notes on American geology, *Amer. Jour. Sci., 35*, pp. 237–251.

Cox, A., R.R. Doell, and G.B. Dalrymple, 1963, Geomagnetic polarity epochs and Pleistocene geochronometry, *Nature, Lond., 198*, pp. 1049–1051. This source was extensively used in writing Chapter 13.

Cox A., R.R. Doell, and G.B. Dalrymple, 1964, Reversals of the earth's magnetic field, *Science, 144*, pp. 1537–1543.

Croll, J., 1864, On the physical cause of the change of climate during geological epochs, *Philosophical Magazine, 28*, pp. 121–137.

Croll, J., 1865, On the physical cause of the submergence of the land during the glacial epoch, *The Reader, 6*, pp. 435–436.

Croll, J., 1867, On the excentricity of the earth's orbit, and its physical relations to the glacial epoch, *Philosophical Magazine*, *33*, pp. 119–131.

Croll, J., 1867, On the change in the obliquity of the ecliptic, its influence on the climate of the polar regions and on the level of the sea, *Philosophical Magazine*, *33*, pp. 426–445.

Croll, J., 1875, *Climate and time*, Appleton & Co., New York. This source was extensively used in writing Chapters 5 and 6.

Dana, J.D., 1894, *Manual of geology*, American Book Co., New York.

Denton, G.H. and W. Karlén, 1973, Holocene climatic variations—their pattern and possible cause, *Quaternary Research*, *3*, pp. 155–205.

Dunbar, C.O., 1960, *Historical geology, 2nd ed.*, John Wiley & Sons, New York. This source was used in writing Chapter 2.

Eberl, B., 1930, *Die Eiszeitfloge im nördlichen Alpenvorlande*, Dr. Benno Filser, Augsburg.

Eddy, J.A., 1977, The case of the missing sunspots, *Scientific American*, *236*, pp. 80–92.

Emiliani, C., 1955, Pleistocene temperatures, *Jour. Geol.*, *63*, pp. 538–578.

Emiliani, C., 1966, Paleotemperature analysis of Caribbean cores P6304–8 and P6304–9 and a generalized temperature curve for the past 425,000 years, *Jour. Geol.*, *74*, pp. 109–126.

Epstein, S., R. Buchsbaum, H. Lowenstam, and H.C. Urey, 1951, Carbonate-water isotopic temperature scale, *Geol. Soc. Amer. Bull.*, *62*, pp. 417–425.

Ericson, D.B., W.S. Broecker, J.L. Kulp, and G. Wollin, 1956, Late-Pleistocene climates and deep-sea sediments, *Science*, *124*, pp. 385–389.

Ericson, D.B., M. Ewing, and G. Wollin, 1963, Pliocene-Pleistocene boundary in deep-sea sediments, *Science*, *139*, pp. 727–737.

Ericson, D.B., M. Ewing, G. Wollin, and B.C. Heezen, 1961, Atlantic deep-sea sediment cores, *Geol. Soc. Amer. Bull.*, *72*, pp. 193–286.

Ericson, D.B. and G. Wollin, 1968, Pleistocene climates and chronology in deep-sea sediments, *Science*, *162*, pp. 1227–1234.

Evernden, J.F., D.E. Savage, G.H. Curtis, and G.T. James, 1964, Potassium-argon dates and the Cenozoic mammalian chronology of North America, *Amer. Jour. Sci.*, *262*, pp. 145–198.

Ewing, M. and W.L. Donn, 1956, A theory of ice ages, *Science*, *123*, pp. 1061–1066.

Fagan, B.M., 1977, *People of the earth*, Little, Brown & Co., Boston.

Fairbridge, R.W., 1961, Convergence of evidence on climatic change and ice ages, *Annals New York Acad. Sci.*, *95*, pp. 542–579.

Flint, R.F., 1965, Deep-sea stratigraphy, *Science*, *149*, pp. 660–661.

Flint, R.F., 1965, Introduction: Historical perspectives, in *The Quaternary of the United States*, (H.E. Wright, Jr. and D.G. Frey, eds.), Princeton Univ. Press, Princeton, pp. 3–11. This source was extensively used in writing Chapter 3.

Flint, R.F., 1971, *Glacial and Quarternary geology*, John Wiley & Sons, New York. This source was extensively used in writing Chapters 1–4 and 9.

Flint, R.F. and M. Rubin, 1955, Radiocarbon dates of pre-Mankato events in eastern and central North America, *Science*, *121*, pp. 649–658.

Forbes, E., 1846, On the connexion between the distribution of the existing fauna and flora of the British Isles, and the geological changes which have affected their area, especially during the epoch of the northern drift, *Great Britain Geol. Survey, Mem.*, *1*, pp. 336–432.

Frenzel, B., 1973, *Climatic fluctuations of the ice age*, (Translated by A.E.M. Nairn), Case Western Reserve Univ. Press, Cleveland and London.

Geikie, A., 1863, On the phenomena of the glacial drift of Scotland, *Geol. Soc. Glasgow, Trans.*, *1*, pp. 1–190.

Geikie, A., 1875, *Life of Sir Roderick I. Murchison*, *2 vols.*, John Murray, London.

Geikie, J., 1874–94, *The great ice age: 1st ed.*, W. Isbister, London, 1874; *2nd ed.*, Daldy, Isbister & Co., London, 1877; *3rd ed.*, Stanford, London, 1894. These sources were extensively used in the preparation of Chapters 3 and 7.

Gilbert G.K., 1890, Lake Bonneville, *U.S. Geological Survey*, *Monograph 1*, U.S. Government Printing Office, Washington, pp. 1–438.

Goldthwait, R.P., A. Dreimanis, J.L. Forsyth, P.F. Carrow, and

G.W. White, 1965, Pleistocene deposits of the Erie Lobe, in *The Quaternary of the United States*, (H.E. Wright, Jr. and D.G. Frey, eds.), Princeton Univ. Press, Princeton, pp. 85–97.

Hansen, B., 1970, The early history of glacial theory in British geology, *Jour. Glaciol.*, *9*, pp. 135–141.

Harrison, C.G.A. and B.M. Funnell, 1964, Relationship of palaeomagnetic reversals and micropalaeontology in two late Cenozoic cores from the Pacific Ocean, *Nature, Lond.*, *204*, p. 566.

Hays, J.D. and W.A. Berggren, 1971, Quaternary boundaries and correlations, in *Micropaleontology of the oceans*, (B.M. Funnell and W.R. Riedel, eds.), Cambridge University Press, pp. 669–691.

Hays, J.D., J. Imbrie, and N.J. Shackleton, 1976, Variations in the earth's orbit: pacemaker of the ice ages, *Science*, *194*, pp. 1121–1132.

Hays, J.D. and N.D. Opdyke, 1967, Antarctic Radiolaria, magnetic reversals and climatic change, *Science*, *158*, pp. 1001–1011.

Heezen, B.C. and M. Ewing, 1952, Turbidity currents and submarine slumps, and the 1929 Grand Banks earthquake, *Amer. Jour. Sci.*, *250*, pp. 849–873.

Hitchcock, E., 1841, First anniversary address before the Association of American Geologists, at their second annual meeting in Philadelphia, April 5, 1841, *Amer. Jour. Sci.*, *41*, pp. 232–275.

Hutton, J., 1795, *Theory of the earth, v. 2*, William Creech, Edinburgh. (Reprinted 1959 in facsimile, Hafner, New York.)

Imbrie, J. and John Z. Imbrie, in prep., Low-frequency components of past and future climatic trends: a model based on the astronomical theory of the ice ages.

Imbrie, J. and N.G. Kipp, 1971, A new micropaleontological method for quantitative paleoclimatology: application to a late Pleistocene Caribbean core, in *Late Cenozoic glacial ages*, (K.K. Turekian, ed.), Yale Univ. Press, New Haven, pp. 71–181.

Irons, J.C., 1896, *Autobiographical sketch of James Croll, with memoir of his life and work*, Edward Stanford, London. This source was extensively used in writing Chapter 6.

Jamieson, T.F., 1865, On the history of the last geological changes in Scotland, *Quart. Jour. Geol. Soc. London*, *21*, pp. 161–195.

Kennett, J.P., 1977, Cenozoic evolution of Antarctic glaciation, the Circum-Antarctic Ocean, and their impact on global paleoceanography, *Jour. Geophys. Res.*, *82*, pp. 3843–3860.

Köppen, W. and A. Wegener, 1924, *Die Klimate der Geologischen Vorzeit*, Gebrüder Borntraeger, Berlin. This source was extensively used in writing Chapter 8.

Kukla, G.J., 1968, *Current Anthropology*, *9*, pp. 37–39.

Kukla, G.J., 1970, Correlation between loesses and deep-sea sediments, *Geol. Fören. Stockholm Förh.*, *92*, pp. 148–180.

Kukla, G.J., 1975, Loess stratigraphy of Central Europe, in *After the Australopithecines*, (K.W. Butzer and G.L. Isaac, eds.), Mouton, The Hague, pp. 99–188. This source was extensively used in writing Chapter 9.

Kullenberg, B., 1947, The piston core sampler, *Svenska Hydro-Biol. Komm. Skrifter*, *S.3*, *Bd. 1*, *Hf.2*, pp. 1–46.

Lamb, H.H., 1966, *The changing climate: selected papers*, Methuen, London.

Lamb, H.H., 1969, Climatic fluctuations, in *World survey of climatology*, *2*, *General climatology*, (H. Flohn, ed.), Elsevier, New York, pp. 173–249. This source was extensively used in writing Chapter 16.

Leverrier, U., 1843–1855, *Connaissance des temps*, 1843; *Annales de l'Observatoire Impérial de Paris*, *II*, 1855.

Libby, W.F., 1952, *Radiocarbon dating*, Univ. Chicago Press, Chicago.

Ludlum, D., 1966, *Early American winters: 1604–1820*, American Meteorological Soc., Boston.

Lurie, E., 1960, *Louis Agassiz: a life in science*, Univ. Chicago Press, Chicago.

Lyell, C., 1830–1833, *Principles of geology*, John Murray, London; *v. 1*, 1830; *v. 2*, 1832; *v. 3*, 1833.

Lyell, C., 1839, *Nouveaux éléments de geologie*, Pitois-Levrault, Paris.

Lyell, C., 1865, *Elements of geology*, John Murray, London. This source was extensively used in writing Chapter 7.

Maclaren, C., 1841, *The glacial theory of Professor Agassiz of Neuchâtel*, The Scotsman Office, Edinburgh, Reprinted, 1842, in *Amer. Jour. Sci.*, *42*, pp. 346–365.

Marcou, J., 1896, *Life, letters, and works of Louis Agassiz*, Macmillan, New York.

Matuyama, M., 1929, On the direction of magnetisation of basalt in Japan, Tyôsen and Manchuria, *Imperial Acad. of Japan Proc.*, *5*, pp. 203–205.

McDougall, I. and D.H. Tarling, 1963, Dating of polarity zones in the Hawaiian Islands, *Nature, Lond.*, *200*, pp. 54–56.

McIntyre, A., W.F. Ruddiman, and R. Jantzen, 1972, Southward penetrations of the North Atlantic polar front: faunal and floral evidence of large-scale surface water mass movements over the past 225,000 years, *Deep-sea Research*, *19*, pp. 61–77.

Mesolella, K.J., R.K. Matthews, W.S. Broecker, and D.L. Thurber, 1969, The astronomical theory of climatic change: Barbados data, *Jour. Geol.*, *77*, pp. 250–274.

Milankovitch, M., 1920, *Théorie mathématique des phénomènes thermiques produits per la radiation solaire*, Gauthier-Villars, Paris.

Milankovitch, M., 1930, Mathematische Klimalehre und astronomische Theorie der Klimaschwankungen, in *Handbuch der Klimatologie*, *I (A)*, (W. Köppen and R. Geiger, eds.), Gebrüder Borntraeger, Berlin, pp. 1–176.

Milankovitch, M., 1936, *Durch ferne Welten und Zeiten*, Koehler und Amalang, Leipzig. This source was extensively used in the preparation of Chapter 8.

Milankovitch, M., 1938, Astronomische Mittel zur Erforschung der erdgeschichtlichen Klimate, *Handbuch der Geophysik*, *9*, (B. Gutenberg, ed.), Berlin, pp. 593–698.

Milankovitch, M., 1941, Kanon der Erdbestrahlung und seine Andwendung auf das Eiszeitenproblem, *Royal Serb. Acad.*, *Spec. Publ.*, *133*, Belgrade, pp. 1–633. English translation published in 1969 by Israel Program for Scientific Translations available from U.S. Dept. Comm. These sources were extensively used in writing Chapter 8.

Milankovitch, M., 1952, Memories, experiences and perceptions from the years 1909–1944, *Serb. Acad. Sci.*, *CXCV*, pp. 1–322 (in Serbo-Croatian).

Milankovitch, M., 1957, Astronomische Theorie der Klimaschwankungen ihr Werdegang und Widerhall, *Serb. Acad. Sci.*, *Mono.*, *280*, pp. 1–58.

Mitchell, J.M., Jr., 1963, On the world-wide pattern of secular temperature change, in *Changes of climate*, Arid Zone Research XX, UNESCO, Paris, pp. 161–181.

Mitchell, J.M., Jr., 1973, The natural breakdown of the present interglacial and its possible intervention by human activities, *Quaternary Research*, *2*, pp. 436–445.

Mitchell, J.M., Jr., 1977, The changing climate, in *Energy and climate*, Studies in Geophysics, National Academy of Sciences, Washington, pp. 51–58. This source was extensively used in writing Chapter 16.

Mitchell, J.M., Jr., 1977, Carbon dioxide and future climate, *Environmental Data Service*, *March*, *U.S. Dept. Comm.*, pp. 3–9. This source was extensively used in writing Chapter 16.

Murray, J., 1895, A summary of the scientific results obtained at the sounding, dredging, and trawling stations of *H.M.S. Challenger*, *Rep. Scient. Res. Voy. H.M.S. Challenger*, *Summary*, *1–2*.

National Academy of Sciences, 1975, *Understanding climatic change: a program for action*, National Academy of Sciences, Washington. This source was extensively used in writing Chapters 4, 16, and Epilogue.

North, F.J., 1942, Paviland Cave, the "Red Lady," the deluge, and William Buckland, *Annals of Science*, *5*, pp. 91–128.

North, F.J., 1943, Centenary of the glacial theory, *Proc. Geol. Assoc.*, *54*, pp. 1–28. This source was extensively used in writing Chapters 1 and 2.

Öpik, E.J., 1952, The ice ages, *Irish Astronomical Jour.*, 2, pp. 71–84.

Penck, A. and E. Brückner, 1909, *Die Alpen im Eiszeitalter*, Tauchnitz, Leipzig.

Phleger, F.B., F.L. Parker, and J.F. Peirson, 1953, North Atlantic foraminifera, *Repts. Swedish Deep-Sea Expedition 1947–1948*, 7, (H. Pettersson, ed.), Elanders, Göteborg, pp. 1–122.

Pilgrim, L., 1904, Versuch einer rechnerischen Behandlung des Eiszeitenproblems, *Jahreshefte für väterlandische Naturkunde in Württemberg*, *60*.

Richtofen, B.F., 1882, On the mode of origin of the loess, *Geological Magazine*, *9*, pp. 293–305.

Ruddiman, W.F. and A. McIntyre, 1976, Northeast Atlantic paleoclimatic changes over the past 600,000 years, in *Investigation of late quaternary paleoceanography and paleoclimatology*, (R.M. Cline and J.D. Hays, eds.), *Geol. Soc. Amer.*, *Mem. 145*, pp. 111–146.

Rutten, M.G. and H. Wensink, 1960, Palaeomagnetic dating, glaciations and the chronology of the Plio-Pleistocene in Iceland, *Int. Geol. Congr. Sess. 21*, *pt. 4*, p. 62.

Sarnthein, M., 1978, Sand deserts during glacial maximum (18,000 Y.B.P.) and climatic optimum (6,000 Y.B.P.), *Nature, Lond.*, *272*, pp. 43–46.

Schaefer, I., 1953, Die donaueiszeitlichen Ablagerungen an Lech und Wertach, *Geologia Bavarica*, *19*, pp. 13–64.

Schott, W., 1935, Die Formaniferen in dem äquatorialen Teil des Atlantischen Ozeans, *Deutsch. Atlant. Exped. Meteor 1925–1927, Wiss., Ergebn. 3*, pp. 43–134.

Shackleton, N., 1967, Oxygen isotope analyses and Pleistocene temperatures re-assessed, *Nature, Lond.*, *215*, pp. 15–17.

Shackleton, N.J. and N.D. Opdyke, 1973, Oxygen isotope and paleomagnetic stratigraphy of equatorial Pacific core V28–238: oxygen isotope temperatures and ice volumes on a $10^5$and 10^6 year scale, *Quaternary Research*, *3*, pp. 39–55.

Soergel, W., 1925, Die Gliederung und absolute Zeitrechnung des Eiszeitalters, *Fortshr. Geol. Palaeont., Berlin*, *13*, pp. 125–251.

Teller, J.D., 1947, *Louis Agassiz, scientist and teacher*, The Ohio State Univ. Press, Columbus.

Urey, H.C., 1947, The thermodynamic properties of isotopic substances, *J. Chem. Soc.*, pp. 562–581.

Van den Heuvel, E.P.J., 1966, On the precession as a cause of Pleistocene variations of the Atlantic Ocean water temperatures, *Geophys. J. R. Astr. Soc.*, *11*, pp. 323–336.

Vernekar, A.D., 1972, Long-period global variations of incoming solar radiation, *Meteorological Monographs*, *12*, Amer. Meteorol. Soc., Boston.

Whittlesey, C., 1868, Depression of the ocean during the ice period, *Proc. Amer. Assoc. Adv. Sci.*, *16*, pp. 92–97.

Wilson, A.T., 1964, Origin of ice ages: an ice shelf theory for Pleistocene glaciation, *Nature, Lond.*, *201*, pp. 147–149.

Wright, H.E., Jr., 1971, Late Quaternary vegetational history of North America, in *Late Cenozoic glacial ages*, (K.K. Turekian, ed.), Yale Univ. Press, New Haven, pp. 425–464.

Zeuner, F.E., *The Pleistocene Period*, Hutchinson, London, 1959. This source was extensively used in writing Chapter 9.

INDEX

Adhémar, Joseph, 195
  flaw in theory of, 75
  proposes astronomical theory of ice
    ages, 69–75
  Revolutions of the Sea, 69
Agassiz, Louis, 19, 21, 22, 25, 26, 33, 35,
    38, 39, 43, 44, 86, 195
  acceptance of his views in America, 45
  British trip (1840), 39
  converts Buckland to glacial theory, 40
  "Discourse of Neuchâtel," 21–22, 26,
    28, 30
  extravagance of his assertions, 44
  portrait, Fig. 5
  professorship at Harvard, 45
  Studies on Glaciers, 30
Albatross expedition, 136
Alembert, Jean le Rond de, 72–73
Alpine terraces, 115, 154, 156
Antarctic ice sheet, 43, 51, 191, Fig. 9
  history of, Fig. 49
Antarctic Ocean, 65
Aphelion, defined, 71
Arctic Ocean, 51, 67
Arrhenius, Gustaf, 126–127, 131, 198
Association of American Geologists, 46
Astronomical theory, *see also*
    Milankovitch theory
  application
    combined with other theories to
      explain the history of climate,
      184–185, 191
    used to forecast climate, 180–181
  historical development
    proposed by Adhémar (1842), 69–73
    flaw in Adhémar's version identified
      (1852), 75
    proposed by Croll (1864), 80–85
    Croll's ideas debated (1864–1894),
      89–96
    proposed by Milankovitch (1924),
      97–111

    Milankovitch controversy
      (1924–1964), 113–122
    Milankovitch revival (1965–1971),
      141–146
    problem of explaining 100,000–year
      cycle recognized (1968),
      158–159
    confirmed (1976), 170–173
Axial precession, *see* Precession
Axial tilt
  changes in, over last 250,000 years,
    Fig. 41
  cycles of, 169
  defined, 69–70
  effect on distribution of sunlight,
    Fig. 25

Ball, Sir Robert, 95
Barbados (West Indies)
  astronomical theory of its sea-level
    history, Fig. 35
  reef terraces on, 143–144
Barnes, John, 142, 198
Bé, Allan, 163
Berger, André, 169, 181
Berggren, W. A., 152, 200
Bernhardi, Reinhard, 21
Biblical flood, 15, 22, 33–36
Bonaparte, Charles, 38
Broecker, W. S., 199, 200
  argues importance of precession cycle,
    144–145
  defines terminations, 157
  finds 100,000-year cycle dominant, 157
  finds reef ages support astronomical
    theory, 143
  uses radiocarbon method to study
    climatic history, 129, 142
Brown University, 139, 143, 158, 162
Brückner, Eduard, 116, 197
Brunhes, Bernard, 147, 197
Brunhes–Matuyama boundary, *see also*

Paleomagnetic time scale
  dated, 149
  defined, 149, Fig. 36, Fig. 39
  identified in deep-sea cores, 149, 151,
    163
Brunhes Normal Epoch, defined, 149,
    Fig. 36, Fig. 39
Buch, Leopold von, 19, 30
Buchsbaum, Ralph, 136
Buckland, William, 35–41, 86, 195
  cartoon of, 41, Fig. 8
  converted to Agassiz' theory, 40
  studies antediluvian hyena den, 35–36
  studies skeleton in Paviland cave, 36–37
  visit to Switzerland, 38–39
Burckle, Lloyd, 163

Calcium–carbonate cycles, 126–127
Cambridge University, 163
Carbon dioxide, effect on climate, 63–64,
    184–186, Fig. 47, Fig. 48
Carboniferous Period
  defined by Lyell, Fig. 21, Fig. 22
  glaciation during, Fig. 49
Catastrophism, 33–35
Catastrophists, 37
Cenozoic Era
  climate decline of, 190, Fig. 49
  defined by Lyell, 90, Fig. 22
  modern definition of, Fig. 23
Challenger expedition, 123–125, 196
Chamberlin, T. C.
  map of North America by, 49, Fig. 10
  reconstructs North American glacial
    sequence, 90, 114
Charpentier, Jean de, 22–28, 195
CLIMAP (project), 201
  core–reconnaissance group, 163
  maps ice-age ocean, 167
  meeting of June, 1972, 164
  organized, 162
  stratigraphic problem recognized,
    162–163
  stratigraphic problem solved, 164
Climate system
  defined, 61
  energy for, 61
  ocean currents, 62
  radiation-feedback effect, 62
  winds, 62
Climatic cycles, Fig. 42, *see also* Climatic

history, Hundred-thousand-year
  climatic cycle last major, Fig. 48,
last major, Fig. 48, Fig. 49
Climatic forecasting
  based on astronomical theory, 180–181
  based on little-ice-age cycle, 183–184
  by considering carbon-dioxide effects,
    184–186
  by duration of past interglacials, 178
  by projecting long-term cooling trend,
    178
  by projecting short-term cooling trend,
    179–180
Climatic forecasts
  of the next 25,000 years, 185–187,
    Fig. 48
  to A.D. 2100, 184–185, Fig. 47
Climatic history
  abrupt change in climate 11,000 years
    ago, 129, 142
  effect of continental drift on, 189–190
  of Paleozoic and Precambrian Eras, 92,
    189–190
  of past billion years, 189, Fig. 49
  of past 100 years, Fig. 44
  of past 1000 years, Fig. 45
  of past 10,000 years, Fig. 43
  of past 150,000 years, Fig. 48
  of past 500,000 years, 189, Fig. 40
  spectrum of, over past 500,000 years,
    169–170, Fig. 42
Climatic optimum, Fig. 43
  defined, 178
Climatic spectrum
  defined, 166
  measured by van den Heuvel, 166
  of past 500,000 years, 169–170, Fig. 42
Cline, R. M., 162
Conrad, Timothy, 45, 195
Continental drift, 104, 189–190
Coral-reef terraces, *see* Reef terraces
Cox, Allan, 148, 199
Croll, James, 77–88, 196
  analyzes axial tilt, 86
  anticipates use of deep-sea records of
    climate, 123
  awarded honorary degree, 87
  calculates orbital eccentricity, 81, Fig. 18
  Climate and Time, 86
  death, 88
  develops astronomical theory, 80–86

develops positive-feedback principle, 83
develops theory of ocean currents, 85
early occupations, 78–79
estimates date of glacial epoch, 85
explains shelly drift, 43
his theory of ice ages debated, 89–96
marries Isabelle MacDonald, 78
photograph of, Fig. 20
theory of ice ages, Fig. 19
The Philosophical Basis of Evolution, 88
The Philosophy of Theism, 80
Curtis, G. H., 148, 199
Cuvier, Baron Georges, 34
Cycles (climatic), *see* Climatic cycles,
    Climatic history

Dalrymple, G. B., 148
Dana, James, 65, 196
    argues against Croll's theory, 93, 95
Darwin, Charles, 37
Dating techniques
    potassium-argon method, 142, 148
    protactinium method, 142
    radiocarbon method, 119–120
    thorium method, 142, 146
Deep-sea cores
    defined, 125
    gravity corer, defined, 125
    Kullenberg piston corer, defined, 126
    Piggot corer, 125
    taken by Lamont Observatory, 128
Deep-sea sediments
    brown clay, defined, 124
    calcium-carbonate (lime) ooze, 126
    diatom ooze, defined, 124
    displaced sediment layers, 127, 128, 129
    foram ooze, defined, 124
    organic oozes, defined, 124
    radiolarian ooze, defined, 124
    studied by *Challenger* expedition, 124
    turbidity-current deposits, defined,
        128–129
Denton, George, 162, 183
DeVries, Hessel, 120
Diatoms, *see* Plankton
Dickson, Geoffrey, 167, 199
Diluvium, 36
    defined, 35
"Discourse of Neuchâtel," *see* Agassiz
Doell, R. R., 148
Donn, William, 66

Dreimanis, Aleksis, 121
Drift
    defined, 37
    diluvial theory of, 37
    formation of, 48–49
    iceberg theory of, 37
    shelly, 43

Earth history, Fig. 49, *see* Geological time
    scale
Earth's orbit
    aphelion, 71
    axial tilt, defined, 69–70
    cardinal points, defined, 71
    eccentricity of, 81
    elliptical shape, 69
    equinoxes, 71
    foci, 70
    perihelion, 70
    precession, 72–73
    rotation of, 72
    summer solstice, 71
    winter solstice, 71
Eberl, Barthel, 117, 197
    tests Milankovitch theory, Fig. 30
Eccentricity (of earth's orbit)
    calculated by Leverrier, 81
    changes in, over past 250,000 years, Fig. 41
    cycle of, 169
    defined, 81, Fig. 17
    emphasized in Croll's ice-age theory, 83–84
    graph of changing values calculated by
        Croll, Fig. 18
Elie de Beaumont, Jean Baptiste, 19, 30
Ellipse, of varying eccentricities, Fig. 17
Emiliani, Cesare, 198, 199, *see also*
        Ericson-Emiliani controversy,
        Oxygen-isotope stages
    defines oxygen-isotope stages, 131–133
    develops oxygen-isotope method, 130,
        136–137
    disagrees with time scale of Broecker
        and Ericson, 137
    graduate study at University of
        Chicago, 135
    "Pleistocene Temperatures," 136–137
    supports Milankovitch theory, 137
Emiliani stages, *see* Oxygen-isotope stages
Eocene Period (Epoch)
    climatic history of, Fig. 49
    defined by Lyell, Fig. 21, Fig. 22

modern definition of, Fig. 23
Epstein, Samuel, 135–136, 198
Equinoxes
   autumnal, defined, 71
   dates of, Fig. 14
   defined, 71
   precession of, Fig. 16
   vernal, defined, 71
Ericson, David, 198, 199, *see also*
      Ericson-Emiliani controversy
   at Lamont Geological Observatory, 128
   at Woods Hole Oceanographic
      Institution, 127
   defines *Globorotalia menardii* zones, 131,
      163, Fig. 33
   develops paleoclimatic (*menardii*)
      method, 129
   estimates age of Pliocene-Pleistocene
      boundary, 133
   identifies displaced layers, 128
   The Deep and the Past, 133
   uses paleomagnetic time scale, 151
Ericson-Emiliani controversy
   develops, 130–131, 137
   subject of National Science Foundation
      conference, 137–138
Ericson's *menardii* curve, Fig. 33
Ericson's *menardii* zones, Fig. 33
Ericson's U-V boundary, Fig. 38
Erratic boulders, 24, 28, 38, 49, Fig. 3,
      Fig. 7
   defined, 20
Esmark, Jens, 21
Europe
   conditions during last ice age in, 11, 14,
      153
   ice-age succession in, Fig. 29
Evernden, J. F., 148, 199
Ewing, Maurice, 66, 128, 133, 136, 149

Fairbridge, R. W., 141
Fermi, Enrico, 135
Filter analysis, 172
Flint, R. F., 120, 137
Florida State University, 168
Foraminifera, *see* Plankton
Forams (bottom-dwelling), 163–164
Forams (planktonic) *see* Plankton
Forbes, Edward, 41, 90, 152
Forecast (climatic), *see* Climatic forecasting
Foster, John, 151

Funnell, Brian, 149, 199

Gauss Normal Epoch, Fig. 36
Geikie, Archibald, 57, 86, 196
Geikie, James, 196
   argues against crustal-movement
      theory, 65, 89
   supports Croll's ice-age theory, 89–90, 95
   The Great Ice Age, 89
Geological time scale, *see also*
      Pliocene-Pleistocene boundary
   Cenozoic, table of, Fig. 23
   Lyell's diagrammatic illustration of,
      Fig. 21
   Lyell's subdivision of, 90, Fig. 22
Gilbert, Grove, 56, 196
Glacial Ages
   late Cenozoic, 189, Fig. 49
   late Precambrian, 189, Fig. 49
   Permo-Carboniferous, 189, Fig. 49
Glacial deposits, *see* Drift, Glacial moraine,
      Glacial till, Outwash deposits
Glacial drift, *see* Drift
Glacial Epoch, in Croll's theory, Fig. 19
Glacial margin
   during ice age, 49
   fluctuations of, in North America, Fig. 31
Glacial moraine, 24
   defined, 14
   on Cape Ann, Mass., Fig. 2
   terminal, 49
Glacial outwash (deposits), defined, 48
Glacial scratches and grooves, 37, 49
Glacial theory, 19–31, 33–46
   resistance to, 41–44
Glacial till(s)
   ablation, 48
   defined, 48
   in Scotland, Fig. 12
   lodgement, 48
   multiple, 56–57
Glacier(s)
   action of, 47–48
   bedrock polished by, 28, Fig. 6
   crevasses, 49
   equilibrium, 48
   Flesch, Switzerland, 24
   meltwater streams, 48
   snow budget of, 48
   valley, 48
Glass, Billy, 151

Globorotalia menardii (foram)
  concentration of, used by Schott as
    climatic record, 126
  curve (*menardii* curve), defined, 131
  defined, 126
  Ericson's zones defined, 131
  illustrated, Fig. 32
  zones of, used by Ericson as climatic
    record, 129–131
Goldthwait, R. P., 121
Graham, John, 149
Gravity core, *see* Deep-sea cores
Great Salt Lake (Utah), 55–56
Greenhouse effect (of carbon dioxide), 64
Greenland ice sheet, 14, 43
Gressly, Amanz, 28
Gulf Stream, *see* Ocean currents
Günz ice age, Fig. 24, Fig. 26, Fig. 29,
  Fig. 30
  defined, 116

Harrison, Christopher, 149, 199
Hays, James D., 151, 152, 162, 163, 167,
  168, 172, 199, 200, 201
Heath, Ross, 162
Heezen, B. C., 128
Hitchcock, Edward, 46
Holocene Epoch
  defined, 91, 178, Fig. 23
  in North America, Fig. 28
Howarth, Henry, 46
Humboldt, Alexander von, 31, 75
Hundred–thousand–year climatic cycle
  explained by Mesolella and Kukla,
    158–159
  found to dominate late Pleistocene
    climatic history, 153–159, Fig. 38
  identified by Ruddiman and McIntyre,
    158
  measured in deep-sea cores, 170
  not accounted for in Milankovitch
    theory, 158
  recorded in Caribbean core V12-122,
    157, Fig. 38
Hutson, William, 162
Hutton, James, 21

Ice age(s), *see also* Little Ice Age, Theories
  of the ice ages
  conditions in Europe during, 153

defined, 11, 177–178
early ideas on, 21–22
geological exploration of last, 47–57
of remote times, 189–190, *see* Glacial
  ages
succession according to Ericson and
  Emiliani, Fig. 33
term coined by Schimper, 28
theoretical succession of European,
  table of, Fig. 29
theoretical succession of North
  American, table of, Fig. 28
Iceberg theory (of glacial drift), *see* Lyell
Ice-raft theory (of glacial drift), *see* Lyell
  objections to, 37–38
Ice sheet(s)
  Laurentide, 51
  mechanics of, 48
  North American, extent during ice age,
    49, Fig. 10
  North American, fluctuations between
    Indiana and Quebec, 121–122,
    Fig. 31
  spreading centers, 51
  thickness of, 52
IDOE, identified, 162
Illinoian ice age, defined, 114, Fig. 28
Imbrie, John, 137–140, 161, 162, 169,
  170, 172, 181, 200, 201
Imbrie, J. Z., 181
Interglacial Age (Epoch), *see*
  Super-interglacial
  defined, 178
  in Croll's theory, Fig. 19
  last, Fig. 48
Irons, James, 79
Isotope stages, *see* Oxygen-isotope stages

Jamieson, Thomas, 53, 196
Jaramillo Event, defined, 149, Fig. 36
Jura (mountains), 19, 20, 24, 30

Karlén, Wibjörn, 183
Kepler, Johann, 69
Khramov, A. N., 148
Kipp, Nilva, 139, 200
Köppen, Wladimir, 103, 197
  advises Milankovitch, 104–105
  compares Milankovitch radiation curves
    with climatic history of Penck and

Brückner, 105, 116–117
Kuhn, Bernard, 21
Kukla, George, 199, 200, 201
    analyzes climatic cycles in Central
        Europe, 153–154
    defines marklines, 154
    disproves Penck-Brückner climatic
        scheme, 156
    finds magnetic reversals in European
        soils, 154
    identifies main pulsebeat of climate,
        154
    interest in loess and soil, 153
    joins CLIMAP project, 162
Kullenberg, Björe, 126, 198
Kullenberg core, see Deep-sea cores
Kulp, J. L., 129

Ladd, John, 163, 200
Lake Bonneville (Utah), 56
    shorelines, Fig. 11
Lake Superior, shorelines, 54
Lamb, H. H., 183
Lamont Geological Observatory
    (Lamont-Doherty Geological
    Observatory), 128, 129, 137, 149,
    156, 162
Lamont, Thomas, 128
Laurentide ice sheet, 51
Leverett, Frank, 114
Leverrier, Urbain, 195–196
    calculates changes in orbital eccentricity,
        80–81
Libby, Willard, 119–120, 135, 198
Lidz, Louis, 138
Little Ice Age, 181, 183, Fig. 43, Fig. 45,
    Fig. 46
    cycle, 183–184
    in Europe, 181, 183
    in New England, 183
Loess (deposits)
    defined, 54
    explained, 54–55
    in Austria, 153
    in Czechoslovakian brickyard, 153,
        Fig. 37
    in North America, 55
Lowell, John, 45
Lowenstam, Heinz, 136
Ložek, Vojen, 153, 199
Ludlum, D. M., 183

Lyell, Charles, 35, 37–38, 39–40, 86, 94,
    152
    ice-raft theory, 22, 25, 37
    Principles of Geology, 37
    proposes crustal-movement theory of
        ice ages, 65
    succession of fossiliferous strata
        according to, Fig. 21

Maclaren, Charles, 51, 195
Magnetic polarity epochs, see
    Paleomagnetic time scale
Magnetism, earth, see Paleomagnetic time
    scale
Marklines, see Soil deposits
    correlated with terminations, 158
Mather, Cotton, 34
Matthews, Robley, 199, 200
    finds reef ages support astronomical
        theory, 144–145
    joins CLIMAP, 162
    proves Barbados reef terraces are
        sea-level records, 143–144
Matuyama, Motonori, 147–148, 197
Matuyama Reversed Epoch, Fig. 36,
    Fig. 39
    defined, 148
McDougall, Ian, 148
McGee, W. J., 57
McIntyre, Andrew, 158, 161, 162
Menardii curve (zones)
    climatic interpretation questioned, 133
    defined, 131
Mesolella, Kenneth, 144, 158, 199
Mesozoic Era, Fig. 49
    in Lyell's classification, Fig. 22
Mid-Atlantic Ridge, 128
Milankovitch, Milutin, 96, 197–198
    calculates ice-sheet response, 108–109
    calculates radiation curve for latitude
        65° N, 105, Fig. 24
    calculates radiation curves for different
        latitudes, 106, Fig. 26
    Canon of Insolation and the Ice Age
        Problem, 109
    death, 111
    decision to attack ice-age problem, 98
    describes mathematical climate of Mars
        and Venus, 103
    discussions with Köppen and Wegener,
        103–106

obtains Ph.D., 98
portrait of, Fig. 27
prisoner of war, 102–103
professorship at University of Belgrade, 98
publishes first treatise on ice-age problem, 101–102
publishes treatise on solar radiation, 103
separates climatic effects of tilt and precession, 107–108
serves in First Balkan War, 101
theory of ice ages, tested by Eberl, Fig. 30
Through Distant Worlds and Times, 109
use of Newton's laws of radiation, 100–101
use of Pilgrim's orbital calculations, 100
Milankovitch theory, *see also* Astronomical theory
application, 117
historical development
   outline of idea first published (1914), 101
   radiation curve first published (1924), 106, 113
   radiation chronology compared with Penck and Brückner curve (1924–1940), 105, 106, 116
   early debate over (1924–1965), 113–122
   detailed version first published (1930), 106
   revised version published (1938), 108
   summary published (1941), 109
   meteorological arguments against (1952), 119
   Emiliani's evidence for (1955), 137
   radiocarbon evidence against (1955–1965), 120–122
   majority opinion against (1955–1965), 122
   Ericson's evidence for (1956–1968), 131
   Fairbridge's evidence for (1961), 141
   interest revived in (1965–1969), 141–146
   supported by sea-level records (1965–1969), 143, 144–145
   failure to account for 100,000-year cycle identified (1968), 158–159
   collapse of Köppen and Wegener's argument for (1968–1975), 156
   modified version confirmed (1976), 169–173
Mindel ice age, Fig. 24, Fig. 26, Fig. 29, Fig. 30
   defined, 116
Miocene Period (Epoch)
   climatic history during, Fig. 49
   defined by Lyell, Fig. 21, Fig. 22
   modern definition of, Fig. 23
Mitchell, J. M., Jr., 179–180, 184–186
Moore, T. C., Jr., 162
Moraine, *see* Glacial moraine
Mount Monadnock (New Hampshire), 52
Multiple-factor method, 139–140, 161
Multiple glaciation, 56–57, 90
Murchison, Roderick, 40
Murray, John, 124

National Science Foundation, 137, 162
Neanderthal people, 153
Nebraskan ice age, defined, 114, Fig. 28
Neuchâtel (Switzerland), 19–20, 28, 30, 38–39
Newberry, John, 57
New Guinea, reef terraces on, Fig. 34
Niagara Falls, evidence on length of postglacial time, 94
Normal polarity epochs, *see* Paleomagnetic time scale
North America
   during ice age, 11, 14, 49
   ice-age map of, Fig. 10
   ice-age succession in, 114, Fig. 28

Obliquity of ecliptic, *see* Axial tilt
Ocean currents
   Equatorial Current in Croll's ice-age theory, 85–86
   explained by Croll, 85
   Gulf Stream, 62, 158
   Kuroshio Current, 62
   role in climate system, 62
Olduvai Event
   dated, 149
   defined, 149, Fig. 36
Oligocene Epoch
   climatic history during, Fig. 49
   modern definition of, Fig. 23

Oozes, *see* Deep-sea sediments
Opdyke, N. D., 149–151, 162, 163, 200,
    201
Orbital history, *see* Axial tilt, Eccentricity,
    Precession
    of past 250,000 years, Fig. 41
Oregon State University, 162
Organic oozes, *see* Deep-sea sediments
Outwash deposits of glaciers, 48
Oxygen-isotope curve, *see* Oxygen-isotope
    stages
    in core A179-4, Fig. 33
    in core V12-122, Fig. 38
    in core V28-238, Fig. 39
Oxygen-isotope method, *see*
        Oxygen-isotope stages, Oxygen
        isotopes
    developed by Emiliani, 136–137
    originated by Urey, 135
Oxygen-isotope stages, *see also* Oxygen
        isotopes
    as stratigraphic tool, 163–164
    in core A179-4, Fig. 33
    in core VI2-I22, Fig. 38
    in core V28-238, 164, Fig. 39
Oxygen isotopes
    defined, 135
    oxygen-16, 135
    oxygen-18, 135

Paleomagnetic time scale
    as a correlation method, 148
    Brunhes Epoch, defined, 148
    Brunhes-Matuyama boundary, defined,
        149
    epochs of normal polarity, defined, 148
    epochs of reversed polarity, defined,
        147, 148
    field-reversal hypothesis confirmed, 148
    Jaramillo Normal Event, defined, 149,
        Fig. 36
    magnetic history of earth, Fig. 36
    Matuyama Epoch, defined, 148
    Olduvai Normal Event, defined, 149
    "paleomagnetic revolution," 151
    recorded in deep-sea sediments,
        149–151
    self-reversal, 148
    used to date Pliocene-Pleistocene
        boundary, 152

Paleozoic Era, in Lyell's classification,
    Fig. 22
Parker, F. L., 127
Paviland (Wales), 36
Peirson, J. F., 127
Penck, Albrecht, 114–116, 197
Penck and Brückner's ice-age succession
    650,000 years long, 116
    climate curve, 116, Fig. 29
    compared with Milankovitch theory,
        105, 116–117
    defined, 114–116, Fig. 29
    disproved, 154, 156
    Great Interglacial of, 116, 131
Perihelion, defined, 70
Permian Period
    defined by Lyell, Fig. 21, Fig. 22
    glaciation during, Fig. 49
Permo-Carboniferous glacial age,
    189–190, Fig. 49
Perraudin, Jean-Pierre, 22–24, 195
Pettersson, Hans, 126–127, 136
Phleger, Fred, 127, 198
Piggot, C. S., 125
Pilgrim, Ludwig, 100, 197
Pisias, N. G., 170
Piston core, *see* Deep-sea cores
Plankton
    coccoliths, defined, 161
    defined, 124
    diatoms, defined, 124
    discoasters, defined, 133
    forams (foraminifera), defined, 124
    Globorotalia menardii, 126, 129, Fig. 32
    radiolaria, 124
Planktonic forams, *see* Forams
Pleistocene Period (Epoch)
    chronology of, 147–152
    defined by international committee, 152
    defined by Lyell, 91, 152, Fig. 22
    in North America, Fig. 28
    modern definition of, Fig. 23
    redefined by Forbes, 90, 152
Pliocene Period, Fig. 30
    climatic history during, Fig. 49
    defined by Lyell, Fig. 21, Fig. 22
    modern definition of, Fig. 23
Pliocene-Pleistocene boundary
    age estimated by Ericson, 133
    dated paleomagnetically, 152
    defined by Forbes, 152

defined by international committee, 152
defined by Lyell, 152
marked by discoaster extinction, 133
Polarity epochs, *see* Paleomagnetic time
scale
Postglacial age, Fig. 28
Post-Pliocene Period, defined by Lyell,
Fig. 21, Fig. 22
Potassium-argon method, *see* Dating
techniques
Precambrian Era
Glacial Ages in, Fig. 49
in Lyell's classification, Fig. 22
Precession
axial precession, defined, 72
changes in, over past 250,000 years,
Fig. 41
cycle, defined, 73
cycles of, measured, 169
of the earth, Fig. 15
of the equinoxes, defined, 72
used in Croll's ice-age theory, 83–85
Prell, Warren, 162
Princeton University, 162

Quaternary Period, Fig. 49
definition of, 92, Fig. 23

Radiation budget, 62
Radiation curves
as computed by Milankovitch for 65° N,
105, 106, 113, Fig. 24
as computed by Milankovitch for
different latitudes, Fig. 26
compared with sea-level records, 143
Radiation-feedback effect, defined, 62
Radiocarbon-dating method, 119–120, 142
Radiolaria, *see* Plankton
in Antarctic cores, 151
RC11-120 (deep-sea core)
climatic record in, 167–168, Fig. 40
raised by Dickson, 167
Recent Period
defined by Lyell, Fig. 21, Fig. 22
definition of, 91
Reef terraces
defined, 143
on Bahama Islands, 143–144
on Barbados, 143–144, Fig. 35
on Eniwetok, 142

on Florida Keys, 142, 144
on Hawaiian Islands, 145
on New Guinea, 145, Fig. 34
Reversed polarity epochs, *see*
Paleomagnetic time scale
Richthofen, Ferdinand von, 54–55, 196
Riss ice age, Fig. 24, Fig. 26, Fig. 29,
Fig. 30
defined, 116
Rubin, Meyer, 120
Ruddiman, William, 158, 200
Rutten, Martin, 148

Saito, Tsunemasa, 163
Sarnthein, Michael, 179
Schaefer, Ingo, 119, 198
Scheuchzer, Johan, 34
Schimper, Karl, 28, 30
Schott, Wolfgang, 139, 197
Scripps Institution of Oceanography, 126,
127, 149
Sea-level history, *see* Reef terraces
application of thorium method to,
142–143
Barbados record of, Fig. 35
New Guinea record of, Fig. 34
of last ice age, 11, 52–54
Seasons
defined, 70
march of, Fig. 13
Schackleton, Nicholas, 140, 162, 163–164,
172, 199, 201
Shelly drift, 43
Soergel, Wolfgang, 117
Soil deposits
in Austria, 153
in Czechoslovakian brickyard, 153, Fig. 37
in North America, 57
marklines, defined, 154
Solstices
dates of, Fig. 14
defined, 71
summer, defined, 71
winter, defined, 71
Sopwith, Thomas, cartoon of Buckland,
Fig. 8
Spectral analysis, *see* Climatic spectrum
defined, 166
used by CLIMAP investigators, 169–170
Spectrum, *see* Climatic spectrum

Stages, isotopic, *see* Oxygen-isotope stages
Stansbury, Howard, 55
Stone Age people, 14, 53, 153
Stratigraphic zones
    based on isotopic stages, 164
    basd on *menardii* curves, 162–163
    defined, 162
Suess, Hans, 120
Sunspots, 63
Super-interglacial, 185, Fig. 48
Swiss Society of Natural Sciences, 19, 24

Talwani, Manik, 149
Tarling, D. H., 148
Terminations, oxygen-isotope, Fig. 38
    correlated with marklines, 158
    defined, 157
Tertiary Period, modern definition of,
    Fig. 23
Theories of the ice ages
    Antarctic ice-sheet-surge theory of
        Wilson, 65–66
    Arctic Ocean sea-ice theory of Ewing
        and Donn, 66-67
    astronomical theory, *see* Astronomical
        theory
    carbon dioxide, 63–64
    crustal-movement theory of Lyell, 65
    dust particles in space, 63
    solar, 62–63
    stochastic, 67
    volcanic dust, 64–65
Thomson, C. W., 123
Thorium method, *see* Dating techniques
Till(s), *see* Glacial till
Tilt, *see* Axial tilt
Tundra, 11, 14
Turbidity current
    defined, 129
    of Grand Banks earthquake, 128–129
Turekian, Karl, 137

United States Geological Survey, 47, 148
University of California (Berkeley), 148
University of Chicago, 130, 135–136
University of Maine, 162
Urey, Harold, 135–136, 198
U–V boundary used to establish

100,000-year cycle, 156
U–Zone of Ericson
    defined, 131

V12-122 (deep-sea core), 139, 156
    100,000-year climatic cycle in, Fig. 38
V28-238 (deep-sea core)
    raised by John Ladd, 163
    "Rosetta Stone" of late Pleistocene
        climate, 163–164, Fig. 39
Val de Bagnes (Switzerland), 22
Van den Heuvel, E. P. J., 166
Van Donk, Jan, 139, 156, 162, 200
Vema (research vessel), 139
Venetz, Ignace, 22, 24, 195
Vernekar, A. D., 169, 201
V–Zone of Ericson, Fig. 33
    defined, 131

Watkins, Norman, 168, 200
Wegener, Alfred, 104, 106, 116–117, 197
Whittlesey, Charles, 52
Wilson, A. T., 65–66
Winchell, Newton, 94
Winds
    role in climate system, 62
    trade, 62
Wisconsin ice age, defined, 114, Fig. 28
Wollin, Goesta, 129, 133, 199
Woods Hole Oceanographic Institution,
    127, 152
Worthen, Amos, 57, 196
Würm ice age, Fig. 24, Fig. 26, Fig. 29,
    Fig. 30
    defined, 116
W–Zone of Ericson, Fig. 33

X–Zone of Ericson, Fig. 33
    defined, 131

Y–Zone of Ericson, Fig. 33
    defined, 131

Zeuner, Frederick, 117
Z–Zone of Ericson, Fig. 33
    defined, 131